STUDIES IN ....STORY

This series, specially commissioned by the Economic History Society, provides a guide to the current interpretations of the key themes of economic and social history in which advances have recently been made or in which there has been significant debate.

Originally entitled 'Studies in Economic History', in 1974 the series had its scope extended to include topics in social history, and the new series title, 'Studies in Economic and Social History', signalises this development.

The series gives readers access to the best work done, helps them to draw their own conclusions in major fields of study, and by means of the critical bibliography in each book guides them in the selection of further reading. The aim is to provide a springboard to further work rather than a set of pre-packaged conclusions or short-cuts.

ECONOMIC HISTORY SOCIETY

The Economic History Society, which numbers around 3000 members, publishes the *Economic History Review* four times a year (free to members) and holds an annual conference. Enquiries about membership should be addressed to the Assistant Secretary, Economic History Society, PO Box 70, Kingswood, Bristol BS15 5FB. Full-time students may join at special rates.

STUDIES IN ECONOMIC AND SOCIAL HISTORY

Edited for the Economic History Society by Michael Sanderson

PUBLISHED TITLES INCLUDE

B. W. E. Alford British Economic Performance since 1945
B. W. E. Alford Depression and Recovery? British Economic Growth, 1918–1939
J. L. Anderson Explaining Long-Term Economic Change
Michael Anderson Approaches to the History of the Western Family, 1500–1914
R. D. Anderson Universities and Elites in Britain since 1800
Dudley Baines Emigration from Europe, 1815–1930
Theo Baker and Dorian Gerhold The Rise and Rise of Road Transport, 1770–1990
P. J. Cain Economic Foundations of British Overseas Expansion, 1815–1914
S. D. Chapman The Cotton Industry in the Industrial Revolution

The Rise and Decline of the British Motor Industry

Prepared for
The Economic History Society by

ROY CHURCH

Professor of Economic
and Social History,
University of East Anglia, Norwich

MACMILLAN

First published 1994 by
THE MACMILLAN PRESS LTD
Houndmills, Basingstoke, Hampshire RG21 2XS
and London
Companies and representatives
throughout the world

ISBN 0–333–45317–4

A catalogue record for this book is available
from the British Library

Printed in China

Series Standing Order

If you would like to receive future titles in this series as they are
published, you can make use of our standing order facility. To place a
standing order please contact your bookseller or, in case of difficulty,
write to us at the address below with your name and address and the
name of the series. Please state with which title you wish to begin your
standing order. (If you live outside the United Kingdom we may not
have the rights for your area, in which case we will forward your order
to the publisher concerned.)

Customer Services Department, Macmillan Distribution Ltd
Houndmills, Basingstoke, Hampshire, RG21 2XS, England.

Contents

Editor's Preface	vii
Acknowledgements	viii
Themes	ix

1 The Origins of British Pre-eminence in Europe | 1
  (i) The rise of the British motor industry before 1914 | 1
  (ii) War and its aftermath: gains and losses | 7
  (iii) The framework of protection: demand at home and overseas | 11
  (iv) Fordism and the British system of mass production: technology and labour | 20
  (v) The dynamics and limitations of 'personal capitalism' | 26
  (vi) Fordism and the British approach to markets and marketing | 32
  (vii) Debilitating environment: structures and strategies | 39

2 The Roots of Decline | 43
  (i) Post-war pre-eminence: attainment and erosion | 43
  (ii) Private investment and public policies: government and industry | 50
  (iii) Manufacturing systems, management and labour | 60
  (iv) The role of organized labour: strikes and productivity | 65
  (v) Industrial relations: Fordism and post-Fordism | 68
  (vi) Fordist structure and strategy: the managerial organization | 71
  (vii) Industrial structure, organization and corporate culture: the origins and performance of BMC | 75

3 The Vicissitudes and Collapse of a 'National Champion' | 84
  (i) Anatomy of a merger: the rise of British Leyland | 84
  (ii) The effects of merger | 87
  (iii) British Leyland's productivity dilemma: markets and productivity | 92
  (iv) The nationalized champion: policies and personalities | 99
  (v) From nationalization to privatization | 103
  (vi) Globalization and the role of multinationals | 107
  (vii) Explaining decline | 115

  Bibliography | 125

  Index | 138

Editor's Preface

In recent times economic and social history has been one of the most flourishing areas of historical study. This has mirrored the increasing relevance of the economic and social sciences both in a student's choice of career and in forming a society at large more aware of the importance of these issues in their everyday lives. Moreover specialist interests in business, agricultural and welfare history, for example, have themselves burgeoned and there has been an increasing interest in foreign economic development and Britain's role in the wider world. Stimulating as these scholarly developments have been for the specialist, the rapid advance of the subject and the quantity of new publications make it difficult for the reader to gain an overview of particular topics, let alone the whole field.

This series is intended for undergraduates, sixth-formers and their teachers. It is designed to introduce them to fresh topics in the curriculum and to enable them to keep abreast of recent writing and debates. Each book is written by a recognised authority and authors are encouraged to present their subjects concisely to a wider readership without being either partisan or blandly disinterested. The aim is to survey the current state of scholarship, not to provide – in the words of the first editor – 'a set of pre-packaged conclusions'.

Studies in Economic and Social History has been successively edited since its inception in 1968 by Professors M. W. Flinn, T. C. Smout and L. A. Clarkson and has expanded with the growth of the subject. In particular there is now a concern to include more social history and more non-English themes. The editor both commissions new titles and receives unsolicited proposals. In this way the series will continue to reflect and shape the ongoing development of this rich seam of history.

MICHAEL SANDERSON
General Editor

University of East Anglia

Acknowledgements

I wish to record the generosity of Professor Krish Bhaskar in allowing me access to the library resources of the Motor Industry Research Unit, of which he is director. I am grateful, too, to Dr W. Mathew for reading a preliminary draft, on which he made critical and constructive comments, to Dr R. J. Overy, for suggestions after reading the penultimate draft, and to Dr Michael Sanderson, whose careful scrutiny of the manuscript was a further source of improvement. The usual disclaimers apply.

R.A.C

The author and publishers are grateful to the Society of Motor Manufacturers and Traders for permission to reproduce Table 9, to the Motor Industry Research Unit for Tables 9 and 10, and to Blackwell Publishers for Table 6.

Every effort has been made to contact all the copyright holders, but if any have been inadvertently omitted the publishers will be pleased to make the necessary arrangement at the earliest opportunity.

Themes

Among the major manufacturing countries before 1914 Britain was the last to establish a sizeable motor industry; it was also the first to witness its collapse as an independent national enterprise. Whereas until the second decade of the twentieth century the emergence of the industry was relatively slow, production overtook that of other European countries between the wars. For a time the exceptionally favourable conditions immediately following the Second World War perpetuated Britain's lead in Europe as the world's second largest motor manufacturing nation and the biggest exporter of cars and of commercial vehicles.

Even before that time the industry's capacity to generate demand for materials and intermediate inputs from other industries, thereby increasing employment, signalled its potential to become a major force in the economy. After the Second World War the industry's strategic importance to the economy was underlined by its capacity to contribute massively to Britain's balance of trade at a time when foreign, particularly dollar, earnings were vital to the economy. This phase in the industry's development proved to be transitory, for the mid 1960s saw the beginning of a precipitous decline. Britain's 10 per cent share in the car output of the major vehicle-producing countries on the Continent of Europe, in the US and Japan had fallen to half that twenty years later. The American multinational companies (MNEs), Ford and General Motors (through its Vauxhall subsidiary), produced a similar volume of output in Britain to that made by British firms, and dominated sales in the domestic market.

Concern for the adverse impact of decline of an industry described by an influential parliamentary committee as of 'central significance' to the British economy led to *de facto* nationalization in 1975. Thirteen years later, the 'national champion', formerly British Leyland/BL, the remaining British-owned mass producer of motor vehicles (in which the Japanese firm, Honda, already held a 20 per cent share), was sold by the government . As 'Rover' the remaining skeleton of the British volume car industry became a subsidiary company of British Aerospace, a mixed defence and property conglomerate. Before the end of 1989 the three surviving British luxury car makers, Jaguar, Aston Martin and Lotus, had been sold to the American multinationals, while the car-making division of Rolls Royce was acquired by Vickers, the major military hardware manufacturer. The remnants of British commercial vehicle (CV) manufacturing were acquired by the Dutch firm, DAF, completing the demise of an independent British motor industry. The speed and scale of the industry's decline is one of the most dramatic developments in Britain's post-war economic history.

Except for war and the immediate post-war periods, the volume of goods and passenger service vehicles produced was roughly one-third the number of private cars and taxis; by value the difference was around one half (PEP, 1950, *Table 3*). During the inter-war period commercial vehicle makers were in many cases separate from car manufacturers, but the latter soon became also the largest CV makers, mainly as producers of lightweight trucks. When in 1968 a small manufacturer of specialist commercial vehicles took over the major British car producer to form British Leyland, the success of the CV branch of the industry was increasingly affected by the new company's performance also as a mass producer of motor cars. This survey, therefore, will concentrate mainly on the car industry, and primarily on British-owned manufacturers. It focuses principally on interpretation rather than narrative. We highlight historians' disagreements and assess the validity of sometimes conflicting explanations for international differences since the establishment of pre-eminence in Europe between the wars and the reasons for decline thereafter.

1   The Origins of British Pre-eminence in Europe

(i)   The rise of the British motor industry before 1914

Among the most striking contrasts presented by the early history of the motor industry is the technical success of French metal manufacturers in exploiting German patent inventions which formed the basis of the motor industry during the 1890s. Another is the scale and rapidity with which the industry was established a few years later in the US where in 1903 production overtook that of France. The emergence of a motor industry in Britain was slow by comparison, and was heavily dependent upon developments and the flow of information, imports and components from the Continent. With a few exceptions, notably the engineer, inventor and entrepreneur, Herbert Austin, who built the first all-British four-wheeled car in 1899/1900, company promoters and speculators showed more interest in the new industry than did the major engineering companies [Saul, 1962; Church, 1979]. While their rationality in this respect has been questioned [Saul, 1962], in part this is explained by the higher rates of return on capital investment which large engineering companies, possessing the financial resources and engineering capacity to make cars in volume, were achieving from other activities (notably the production of armaments [Irving, 1975]). Not until a broad consensus evolved among engineers and public on what constituted dominant motor design did other British engineer-entrepreneurs, many of whom were cycle manufacturers, began to invest in the sizeable production of British-made vehicles [Saul, 1962; Harrison, 1981].

A series of successful company flotations from 1905 reflected in part the resilience of investors following the disturbing and deterrent effects of the activities of financial speculators dealing in the shares of motor and related companies. No fewer than 221 firms entered the industry between 1901 and 1905, of which 90 per cent had either discontinued motor production or ceased trading altogether by 1914 [Saul, 1962, *Table I*]. Following the general liquidity crisis of 1907 business confidence returned (Michie, 1981; Lewchuck, 1985a], a recovery which the stabilization in design and the reduction in risk which that implied for investors may have assisted and strengthened [Nicholson, 1983, *3*; Harrison, 1981]. The main feature of the dominant design was the basic power train, which incorporated engine, transmission, clutch, drive shaft, differential and axle; these were the mechanical components which generated power and transmitted it to the driving wheels attached to the chassis. This standard form superseded the various three- and four-wheeled vehicles which were little more than tri-cars, quadricycles or dog carts. Other dominant design features by this time included column (rather than tiller) steering, front-mounted engine enclosed within an embryo bonnet, seating side-by-side (rather than face-to-face), pneumatic (rather than solid) tyres, and the option of a completely enclosed saloon car. Petrol became the acknowledged preference as the power source [Caunter, 1957].

Slow off the mark, the British industry lagged behind until a sharp rise in production began to close the gap between French and British output. While French production rose by barely one-third between 1909 and 1913 British output increased threefold. By 1913 British car (including commercial vehicle) production had reached 34,000, compared with 45,000 in France, and 23,000 in Germany. The European total, however, was less than a quarter of the output in the US.

Disparity in the size of national production did not mirror precisely the extent of national markets. Ownership density in the US was one vehicle for every 77 people in 1913, a figure derived from 1.26 m in use. The comparable densities in Europe were 165 in Britain, 318 in France, and 950 in Germany [Bardou, Chanaron, Fridenson and Laux, 1982; US Bureau of Census, 1976]. Britain, therefore, was the largest market in Europe, to which the French were the major exporters. Much has been made of Britain's lag behind the Americans in the speed and scale of development. Saul blamed British engineers, whose approach to the market, to product development and to production methods he described as having been 'well nigh fatal' [Saul, 1968, *224*]. Saul condemned the failure to invest in plant to produce in volume a small, inexpensive car of the kind which by 1910 dominated the American market and was more common in France than in Britain [Saul, 1962]. He also criticized the passion shown by British engineers for an irrational pursuit of technical perfection and individuality, the exceptions having been individuals trained abroad. Lack of attention to new production methods, he argued, led to low productivity.

The basic problem is seen to have been the inability of the industry to 'free itself from the older traditions of engineering . . . the most crucial weakness was the failure to realize that the new engineering industries called for a complete change from the old ways of mechanical engineering so as to make full use of the new techniques of production engineering' [Saul, 1968, *224*: 1962]. Specifically, Saul contrasted British methods of manufacture with those based on repetitive production and the assembly of inter-changeable parts. His criticism was that not only were these pro-cesses not widespread before 1914 but that British firms did not organize production in such a way as to exploit the new machinery to the full. In other words, the artisanal craft-based methods of manufacture continued to predominate until 1914 and beyond, whereas the American industry had already been transformed. Interchangeability, advanced division of labour and the assembly of standardized parts along a production line culminated in 1913 with Henry Ford's moving assembly line installed at the Highland Park factory near Detroit.

Finally, underlying these particular production weaknesses was the lack of 'commercial acumen' among those responsible for designing and selling cars [Saul, 1962, *41*]. These conclusions have been echoed elsewhere, contributing to a conventional wis-dom. Mathias, for example, referred to decision-makers having paid more attention to technical than to market criteria, and added the absence of cost consciousness in most firms to the list of the industry's weaknesses [Mathias, 1983], a gloss on Saul's con-clusions which points in the direction of entrepreneurial failure. Are his wide-ranging criticisms valid in the light of subsequent research?

A comparison with the French industry suggests that higher levels of production in France cannot be explained by superior entreprenurial performance or greater engineering imagination. The trend towards integrated manufacture, which has been attrib-uted to the desire of engineers to make the entire vehicle them-selves and to reject standardization, was no less characteristic of French manufacturers [Laux, 1976]. Laux's detailed examination of French manufacturers' approach to markets found little evidence of low costs from large-scale production as supposed by Foreman-Peck [1979]. Lewchuck argued that vertical integration did not, in any case, preclude standardization. Even by 1905 some of the larger British vehicle manufacturers were using American and British machine tools designed for repetitive manufacture and the assembly of virtually interchangeable components [Lewchuck, 1987]. Typically in both countries production occurred in small batches, compared with the sequential flow production system in use in the large American factories. Small batch production involved a division of work between several gangs of workers who moved along a row of stationary assembly stands. Such a system also allowed rectification of defects by hand in product or jig design and fixtures. Whereas Saul took this as evidence of conservatism Lewchuck emphasized the British system's flexibility, suited for factories characteristically producing a variety of models for a limited and socially stratified market [Saul, 1962; Lewchuck, 1987]. Lewchuck has also challenged the blanket condemnations of British compared with American productivity. His comparison of the productivity of British and American manufacturers making similar kinds of vehicles showed little difference, although the estimates which were the basis for the comparison were few and may not have been representative [Lewchuck, 1987].

As for the superiority of French-trained engineers, Laux found that barely one-fifth of the leaders of the French motor industry received a 'high class education' in engineering. It is also evident that their approach to manufacturing methods was similar to that employed by their British counterparts. Furthermore the favourable French balance of trade in motor vehicles cannot be explained by superior marketing, for the large market for French cars in Britain was developed primarily by British agents [Laux, 1976]. The larger British and French firms typically supplied a variety of models at prices which the rich, the professional and business classes could afford. An emphasis on technical design and quality, 'fit and finish', rather than price competition, was characteristic of the motor trade in both countries until shortly before the war [Church, 1981]. Indeed, the absence of striking differences between the strengths and weaknesses of the supply and quality of the factors of production and the extent and characteristics of the market in the two countries suggests that the critical factor explaining the more rapid early development of the French industry may have been a chance competitive advantage. This was secured in 1888 when Gottlieb Daimler approached French metal-manufacturing firms with a view to their becoming the first to manufacture his patent petrol engine, and so initiate an industry based on the internal combustion engine [Nubel, 1987].

After the initial pioneering phase of the motor vehicle, the lag of British and French production behind that in the US and the contrasts in methods and approach cannot be explained without reference to the enormous difference in the size of internal markets. The levels and distribution of real income, and a social geography and rail density which by the automobile age had given Europe close and efficient communication systems both within and between towns and cities, were key differences. They were important factors which shaped entrepreneurs' perceptions of the market and the types of vehicles that could be sold. On the eve of the First World War the high productivity of Ford's American plant was based on economies of scale from production in large volume, interchangeable parts, special-purpose machinery and flow production, combined with a disciplined, highly-paid labour force. In 1913 Ford produced over 200,000 units, compared with some 5000 by Peugeot, the largest French manufacturer, and 3000 by the Wolseley Motor Company, the largest British car maker [Bardou *et al.*, 1982]. The capital investment required to produce on an American scale, however, was justified only if it seemed possible to those in the industry, or to newcomers, that vehicles could be sold in such numbers and at a profit.

Perceptions of market possibilities began to alter both in Britain and in France shortly before the First World War, evidence of which is the repositioning by some manufacturers who began to build cars to sell within a lower price range than hitherto. In Britain the catalyst was Henry Ford who was persuaded by Percival Perry, formerly an importing agent selling Ford cars in England, to establish a branch. Tax advantages explain why in 1911 the branch, opened in 1909, was replaced by the Ford Motor Company (England), wholly-owned by the parent company. The assembly of Model T cars from imported kits began at Trafford Park near Manchester in the same year [Wilkins and Hill, 1964]. The price of the Ford Model T Runabout was £135, and the Tourer £150. Designed to suit American road conditions in rural and urban America, to travel long distances, and to be within the purchase range of farm and business users, both were regarded by the British press as unattractive 'cheap and nasty' vehicles. Built with high horsepower and a slow-speed engine, the Model T could achieve smooth running without requiring the level of technical precision in the machining of parts that was needed in constructing the typically high-speed engines used in British cars [Wilkins and Hill, 1964; Saul, 1962].

The Model T proved successful in the British market because of its low price, roughly 25 per cent cheaper than the Morris Oxford. This was the 'popularly-priced' car introduced in 1913 to compete with Ford by the British car maker, W. R. Morris, whose recently established company was later to become Britain's largest car producer [Overy, 1976]. The high productivity of American parts suppliers incorporated in the knocked-down kits dispatched from Detroit gave Ford an important cost advantage derived from large-scale production for the huge American market. Combined with Ford's highly efficient assembly plant at Trafford Park, Ford virtually created and dominated the cheap market for motor cars before 1914. Estimates of the sale of cars in the price range £200 and below, regarded by contemporaries as below the luxury and semi-luxury threshold, suggest that the 7310 Ford cars sold in 1913 accounted for more than 60 per cent of the total in that price range [Church, 1982].

While a handful of well established, though small, manufacturers ventured into the lower segment of the market from 1912, the major British entrant into the popular car market was a newcomer, W. R. Morris. Like most other car makers his origins were in the cycle trade, although whereas the founders of most firms which survived into the 1930s had some knowledge of engineering, Morris was essentially a mechanic with an innovative inclination combined with a willingness to take risks. In 1912 the newly formed W. R. M. Motors began to prepare for the low-cost volume production of cars aimed at a popular market. The starting capital for this venture was £1000, which was supplemented by financial backing from the Earl of Macclesfield. Morris moved against the trend towards vertical integration by assembling cars entirely from components built by specialist suppliers on contracts, the system widely employed in the cycle industry. This strategy enabled him to exploit the human and physical capital resources of the engineering trades, to take advantage of their economies of scale, and to expand rapidly without the need for large capital expenditure [Andrews and Brunner, 1955; Overy, 1976].

The price of the Morris 8 h.p. Oxford basic model, first sold from a blueprint at the Motor Show in October 1912, was £175, enabling it to compete with the handful of other British cars aimed at the same market made by the Singer, Standard and Hillman Motor companies. Built to conventional high European standards of materials and finish, the Oxford incorporated a multi-cylinder engine of low horsepower, high speed and high efficiency. In order to compete with Ford, however, Morris planned a second model, the Cowley, lower in horsepower than the Oxford and lower in price. To meet his requirements for supplies of low-cost components in large volumes to make possible large-scale, low-cost assembly, Morris turned to the US, but his plan to commence volume production in 1915 was checked when war intervened [Andrews and Brunner, 1955].

(ii) War and its aftermath: gains and losses

Historians disagree on the effects of the First World War on British industry. Some have stressed the stimulus which virtually compelled British firms to adopt the production methods already widespread in the United States. Others have emphasized the damage caused to the economy by postponing the transfer of resources from the production of traditional goods to the manufacture of new products, notably consumer durables, thereby delaying structural change [Richardson, 1965; Alford, 1981].

Effects on the motor industry were both positive and negative. War conditions restricted the demand for cars at a time when Morris was poised for mass production. The McKenna tariff introduced in 1915, imposing a 33 1/3 per cent *ad valorem* duty on cars and components, was intended to limit the import of an item 'extensively used solely for the purpose of luxury' and to save shipping space [Plowden, 1971, *110*]. One effect was to eliminate Morris's potential cost advantage over other British manufacturers by cutting off high-productivity American suppliers. Another was to accelerate the substitution at the Ford factory of parts made in Britain for those hitherto imported from the parent company. When it became clear that the tariff would remain in place after the war Ford's policy from 1920 was to move towards local manufacture entirely. By 1924 Ford was countering a 'Don't buy

7

American' campaign by publicizing that Ford cars assembled at Trafford Park were 92 per cent British built, though at that time the Ford factory at Cork was a major parts supplier [Wilkins and Hill, 1964].

Alford stressed the gains in efficiency resulting from the stimulus war production gave to the adoption of interchangeable parts [Alford, 1986]. The enforced learning experience of munitions manufacture did benefit Morris, but some other larger, longer established car producers already possessed considerable experience of the use of American special purpose machinery, interchangeability, and production with fewer skilled workers. Moreover, those large manufacturers, such as Austin and Wolseley, who supplied aeroplanes, armoured vehicles, ambulances and lorries for the war effort could learn less from the limited production runs normally required for these items. They also found difficulty in applying techniques used to manufacture shells to the production of immensely more complex motor vehicles after the war [Lewchuck, 1987; Church, 1979].

Of critical importance for post-war development was the effect of using standard jigs and tools and the subdivision of processes into simple tasks. This allowed semi-skilled, usually female, labour to be employed in the place of skilled male fitters [Andrews and Brunner, 1955]. While this trend was present in car plants before 1914, war accelerated the progressive dilution of labour. Facilitated by the intervention of the Board of Trade in agreement with the Engineering Employers' Federation and the trade unions, jointly represented on a committee of production, employers were free to introduce or extend dilution for the duration of the war. The learning experience of how machinery might be used to deskill large elements of the production process, at least outside the body shops, was to affect employers' approach to manufacturing methods. It also affected the nature of managerial control after the war. The Pre-war Practices Act envisaged that one year after hostilities ceased job controls hitherto exercised by the various craft unions would revert to managers. In fact the effects of dilution were to prove permanent and fundamental to production and labour management [Cole, 1923].

War contracts allowed Morris not only to more than double the size of his Cowley factory and to install modern machinery financed by government, but also to develop a circle of subcontractors and build up a larger, better trained workforce. The stock of American parts acquired in 1915 permitted car production to continue, albeit at low levels, throughout the war, when many of the technical problems affecting both product and production were solved [Overy, 1976].

While for some firms the enforced learning experience of munitions production proved beneficial, some of the largest car manufacturers suffered adverse effects resulting from the very large expansion of premises and plant not easily converted to peacetime use. Whereas Morris's factory merely assembled parts for munitions, manufacturers such as Austin and Wolseley found themselves with large purpose-built factories which involved heavy capital expenditure and delay in resuming car production after the war. Austin's works manager declared that during the war the size of the workforce at Longbridge rose from 2000 to 22,000. He described the plant as it existed in peacetime conditions in 1918 as 'of an entirely useless character for its needs' [Church, 1979; Lewchuck, 1987]. The evidence suggests, therefore, that even within a single industry war affected different firms in diverse ways.

In the longer term, however, all were affected by the protection offered by the McKenna duties which, with a brief interruption in 1924–5, were retained for nearly 50 years (pp. 11–14). War was also the catalyst which led to a reversal of Ford's dominance of the British market. When the war ended it became clear that Henry Ford wished to operate Trafford Park as no more than a branch, leaving little scope for local initiative. Perry was in poor health following his major wartime role in the Ministry of Munitions, which he had combined with managing the Ford Motor Company (England) while he was also at odds with his pacifist employer, Henry Ford, and other senior managers in Detroit. This was the combination of circumstances which precipitated Perry's forced resignation as managing director. Henry Ford's decision to replace him with American managers was to have serious consequences both for the American company and for the development of the British industry in which in 1913, on a rapidly rising curve, Ford accounted for roughly 20 per cent of the industry's total output [Wilkins and Hill, 1964].

The immediate post-war conditions brought chaos to the vehicle market. There was a decline in private motoring during the war, due partly to the diversion of productive capacity to munitions and partly to restrictions on the importation of cars, except under licence. These, in addition to the tariff, created a pent-up demand [Church, 1979]. The demonstration effects of the military use of vehicles seem to have contributed further to this demand. These influences, to which was added the removal of petrol restrictions, produced a climate of expectancy and optimism among would-be consumers and potential suppliers alike. The mood of the time can be gauged from the crusading motto, coined by the British Motor League, of 'Motoring for the Million' [Plowden, 1971]. Despite the tariff, imports were sucked in mainly from the United States [Miller and Church, 1979]. Soaring car prices stimulated adaptation of the plant of pre-war car-making firms and led to entry into the industry of a host of new producers [Maxcy, 1958].

Erstwhile general engineers, garage proprietors and such unexpected aspirants as the Manchester Cooperative Wholesale Society were among those in 1919 who announced plans for production on a scale reflecting little more than inflated expectations, which quickly proved to be illusory. The *Economist* referred to the many new companies, never having made a single motor vehicle, simply acquiring empty premises, issuing prospectuses illustrated with cars and containing generous profit estimates, whose expectations outran the economic realities of the trade. Between 1919 and 1920, 40 new makes of car were offered on the market [Plowden, 1971], but the principal beneficiaries of the overheated market were the high-productivity producers in the United States which, despite the tariff, were responsible for most of the 34,000 units shipped to Britain during the hectic post-war boom [Miller and Church, 1979].

Many of the new British entrants were unsuccessful and few survived. The expanded capacity and flood of imports sparked off cut-throat competition when depression in the wider economy was transmitted to car sales, bringing about the deepest depression the industry was to experience before the Second World War [Miller and Church, 1979]. This was a critical period, when the industrial structure changed from one of intense competition consisting mainly of a large number of small firms possessing a high mortality rate to one dominated by a few relatively large firms with a comparatively low mortality. The number of car-producing companies fell from 88 in 1922 to 31 in 1929, by which time the industry was dominated by only three, all British, companies: Morris Motors, Austin and Singer, which together accounted for 75 per cent of car production in Britain [Maxcy and Silberston, 1959].

(iii) The framework of protection: demand at home and overseas

Critics of the industry's performance who have compared the scale of British production and density of car ownership with those in the United States have concentrated either on supply factors or those affecting demand. Later we shall consider the determinants of supply, but in sections (iii) and (iv) we shall first discuss the effects on demand of factors external to firms, followed by a consideration of the character and consequences of corporate marketing strategies, particularly the extent to which they may have restricted demand.

The principal determinant of the demand for cars, from first time purchasers, from those replacing a vehicle, and from buyers of second-hand cars, has been the level of consumers' incomes. The second most important factor has been the price of cars in relation to general price levels in the economy [Maxcy and Silberston, 1959]. The pent-up post-war demand for vehicles sucked in imports, presenting a serious threat to British manufacturers in the process of adapting plant to peacetime production. The request by the Society of Motor Manufacturers and Traders (SMMT) that the McKenna duties should be extended, at least until the end of 1920, was formally rejected by the Board of Trade, yet no move was made to remove the duties [Plowden, 1971]. The onset of the industry's worst depression of the interwar period in 1920–1 seems to have been further justification for inactivity on the part of the Board of Trade [Plowden, 1971]. Virtually by accident a temporary measure aimed at checking luxury consumption in wartime became a protective framework which was to have a major effect on the subsequent history of the industry.

The duties were retained, with one brief suspension in 1925, throughout the interwar period, becoming consolidated within the general tariff structure in 1938 and remaining unaltered at the 33 1/3 per cent level until 1956. From 1926 those duties also applied to commercial vehicles. From the standpoint of manufacturing efficiency, the importance of the tariff lies in its possible effect on the size of the market by choking off imports. In a broader context Richardson dismissed the effects on British industries of the general tariff on imports introduced in 1931 as having been insignificant [Richardson, 1961]. However, Capie's calculations of effective protection rates, a concept which gives expression to the margin of protection on value added in the process of production rather than simply on the price of the product, led him to draw a different conclusion. Moreover, his estimates showed that the tariff's effect in reducing import penetration was greater in the case of motor vehicles than for any other product category [Capie, 1983]. An indication of the degree of effective protection enjoyed by the car industry is the jump in the share of net imports in home sales from 15 to 28 per cent during the short-lived experiment in free trade in 1925 [Miller and Church, 1979].

Foreman-Peck emphasized a further effect of the tariff in enabling the industry to achieve lower costs and prices as a result of economies of scale by the large producers, Morris, Austin and Ford [Foreman-Peck, 1981a]. This was undoubtedly true for Morris and Austin, but managerial failure during the 1920s, and later large excess capacity at the new Dagenham plant throughout the 1930s, suggests that scale economies for the American subsidiary were illusory before the Second World War [Wilkins and Hill, 1964]. Between 1924 and 1934 the SMMT index of car prices dropped from 100 to 52; in 1938 the figure was 49, but this represented an absolute price which was 30 per cent above that of the average car in the US [Rhys, 1972].

The tariff contributed to the persistence of this differential, though the effects of the tariff may be interpreted differently in each of the two decades, mainly because of the transition from a stage of 'initial demand', which lasted until the late 1920s, to that of a 'mature market' which developed after the slump. By initial demand is meant the stage when consumers who could afford to buy were persuaded that motor vehicles had reached a level of development at which ownership seemed to be desirable. The stimuli to demand during this phase were product development, publicity and communications, and the widespread practice of instalment buying [Maxcy and Silberston, 1959]. Elasticities of income and price (or motoring costs as a whole) were likely to have been less important in this stage than the elasticity of product improvement – the propensity to reallocate existing incomes to cars as their reliability improved [Maxcy and Silberston, 1959; Miller and Church, 1979]. During this phase the tariff enabled the infant volume car industry to establish itself on a competitive basis within the home market, where manufacturers lowered costs and prices to compete with each other.

It seems doubtful whether price elasticity is central to an explanation for the rapid growth in car sales and production during the twenties. Home sales reached a peak in value in 1925 which was exceeded again only in 1936 and 1937. However, a steep fall in car prices between 1925 and 1929 saw a drop in sales to first time buyers. This suggests that even large real price reductions failed to induce enlarged consumption in the late twenties [Miller and Church, 1979] and that from this period changes in real income became the key to the level of demand. The industry had reached the stage envisaged by W. R. Morris in which ownership would percolate down 'the pyramid of consumption' and new purchases would become less stable [Andrews and Brunner, 1955].

The contrast with growth in the 1930s is obvious; the real average value of new cars purchased between 1932 and 1937 fell by only 8 per cent whereas new owner sales rose more than threefold. This suggests that the minimal fall in prices was more than compensated by the rise of middle-class real incomes after the slump, when the great engine of income elasticity came into operation. At the same time, the dramatic shift in the structure of sales favouring cars of up to 10 h.p. points to consumers' sensitivity to annual running costs, which in the 1930s might be one-third of the original purchase price of a car of 8 h.p. [PEP, 1950; Bowden, 1991] (see Table 1). In the sense that new car sales came to depend

Table 1 *New Car Registrations by Horsepower 1927–38 (percentages)*

	1927	1930	1934	1938
Less than 8 h.p.	16	27	23	30
9 to 12 h.p.	34	26	51	48
13 to 16 h.p.	33	32	17	11
17 and above	16	15	9	11

Source: PEP (1950), Table 17.

primarily on replacement rather than first time purchases, a phase of 'mature demand' was reached in the 1930s [Maxcy and Silberston, 1959; Miller and Church, 1979]. None the less, throughout the 1930s car ownership continued to be a middle-class phenomenon, for the price of even an average family car could be compared with semi-detached house prices in some provincial towns [Bowden, 1991].

It seems possible that by keeping prices higher than they might have been otherwise the tariff checked home demand, preventing prices from falling to levels at which elasticity affected sales to lower income first time buyers. For two reasons this hypothetical relationship is at least open to debate. First, the main imports in the 1920s were American cars with substantially higher running costs than British-built vehicles. Second, the tariff was not the sole measure affording protection.

The effects of taxation in protecting the industry are disputed, as are its consequences for exports and for the level of home demand. Taxation on car ownership and use in Britain was a flat rate annual tax and a duty on petrol, and it is the former which has been the subject of disagreement, both among contemporaries and historians. The horsepower tax, as it became known, took its description from the formula used for the annual tax on cars introduced in the budget of 1909 to raise revenue intended to contribute to the financing of road-building and improvement [Plowden, 1971]. Based on a vehicle's piston diameter multiplied by the number of cylinders in the engine, the initial formula devised by the Royal Automobile Club (RAC) in 1906 was intended originally to guide potential buyers when comparing prices, at a time when the power of the engine was closely related to cylinder capacity. The effect of using the formula for tax purposes, with which the representatives of the motor trade and industry concurred, was to tax high-powered, high-rated cars (according to the equation) more heavily than others. This meant that cars with engines of equal cubic capacity were taxed more heavily if the piston area of their engines was larger.

On its introduction the rates charged rose with horsepower, though not proportionately. In 1920, however, the decision to adopt proportionality increased the progressiveness of the tax, which was fixed at £1 per horsepower. Comparisons between the incidence of tax on British models of different horsepower revealed relatively modest differences, for example adding 1s.2d. per week to the running costs of a 12 h.p. car compared with one of 8 h.p. [Maxcy and Silberston, 1959]. By comparison with the Model T Ford, however, rated at 22.5 h.p. under the RAC formula, the weekly difference was 4s.1d. Whereas before 1920 the Model T paid £6 6s., the new figure was £23 (the same for a Rolls Royce Phantom which was several times more expensive) adding roughly 10 per cent to its 1921 price [Plowden, 1971]. In this way, apparently by historical accident, British cars received an advantage over American imports the specifications of which attracted high rating and tax [Maxcy and Silberston, 1959].

Any suggestion that the tax created the British small, high-performance light car which dominated sales and production in Britain between the wars [Wyatt, 1968] is without foundation. Not only did its origins precede the introduction of the tax, but the typical engine design in France and in Germany (where no vehicle tax was payable) was similar to that in Britain. So too was the structure of sales in those countries. European road conditions, consumer preferences and manufacturers' approach to production and the market, rather than taxation, are the factors which help to explain Anglo-European differences in vehicle design [Plowden, 1971].

It has been argued that in one respect the horsepower tax was ineffective and in two others detrimental to the industry's progress. Bowden concluded that even as a protective measure the tax was a failure. The evidence adduced in support of this view is Ford's 20 per cent share of the British market in 1937 [Bowden, 1991]. This, however, consisted of cars manufactured in Britain after Perry had finally persuaded the Ford management to produce cars designed specifically for the British market and therefore avoid the discriminatory effect in the horsepower tax [Wilkins and Hill, 1964]. It marked the introduction of Ford's first model containing a high-speed engine, small cylinder bore and low horsepower, manufactured in its entirety at the Dagenham plant. It was also a reassertion of British market imperatives by Sir Percival Perry, reinstated in 1928 as the managing director of Ford in England, whose dismal record in Britain since his departure in 1918 he was reappointed to reverse [Wilkins and Hill, 1964].

The short-lived rise in imports from North America following the tax reduction in 1935 suggests that the horsepower tax did add somewhat to the protection afforded by the duties. Imports remained, however, at low levels. In 1937 they represented less than 6 per cent of total car sales in Britain, higher than in previous years since the post-war boom in 1920, and mainly the result of dumping on the British market by Germany and Italy [Miller and Church, 1979]. A highly protected home market was supplemented by Imperial Preference, which from 1938 gave British producers an advantage over their European competitors. In that year 97 per cent of British car output was sold in protected markets, either in Britain or in the Empire. The trade in commercial vehicles reveals a sharp reduction of imported American vehicles beginning in 1926, a consequence of the extension of the McKenna duties to commercial vehicles in that year. From 1930 both Ford and General Motors (through its Vauxhall subsidiary) manufactured CV chassis in entirety in Britain [Miller and Church, 1979]. What is clear from the surge of imports that occurred when the tariff was suspended briefly in 1925 is the inadequacy of the tax alone in providing effective protection from superior American productivity.

Contemporaries criticized the tax not only because of its effect on car prices but for providing a disincentive to the production of large-engined cars of the American type which were popular in potentially important export markets, mainly within the Empire. This was also the conclusion of the National Advisory Council in its 1947 Report, which led to the replacement of the horsepower tax with a tax based on cylinder capacity only [NAC, 1947; Plowden, 1971]. While accepting that the tax did impose some bias towards low-powered cars, Maxcy and Silberston concluded that it was less important as a hindrance to exports than such factors as income and the nature of terrain [Maxcy and Silberston, 1959]. Manufacturers' views differed during the late 1920s. By this time the American import penetration had been checked, partly by the tax and partly by the reintroduction of the duty on petrol in 1928 which raised the running costs of American-built vehicles.

Some manufacturers conceded that their own model designs had been influenced by the tax while others denied its importance [Plowden, 1971]. Engine type in British cars was not the only cause for complaint from Australia, one of Britain's best export markets at that time. Excessive prices, the lack of a standard wheel track and too little ground clearance were added to the criticisms of underpowered engines unsuitable for Australian road conditions [Plowden, 1971]. These complaints strengthen the view that other factors were more important in explaining export performance.

A further criticism of the tax, reduced by 25 per cent in 1935, was that the level of tax reduced home consumption and thereby reduced the possibility for economies of scale in the industry. One calculation has shown that even after the reduction direct tax could represent 36 per cent of the total burden of car taxation, including petrol tax and insurance. The average tax per 1000 c.c. was eight times that in the US, nearly twice that in Germany and 50 per cent greater than in France [Bowden, 1991].

The rise in car sales in the home market which followed the tax cut in 1935 has been interpreted as evidence of the retardative effect of tax-inflated car prices. It is for this reason that Bowden adversely contrasts motor taxation policy in Britain with the various tax concessions introduced by the German government from 1933, which were intended to stimulate car purchase.

Bowden condemned the British government's fiscal stance as having been 'particularly unenlightened' because of the adverse effects on the demand for cars [Bowden, 1991, 258–60]. By contrast the principal aim of the German government's *Motorisierung* policy, of extensive road-building, fiscal concessions and encouragement to enrol in the National Socialist Car Corps, was to stimulate employment [Overy, 1975; Blaich, 1981; Reich, 1990]. Britain, on the other hand, neither experienced comparable depths of depression nor, by European standards, low car density.

Car ownership in Germany almost doubled between 1934 and 1938, although (as was the case in Britain) cars of medium size accounted for most of the increase [Blaich, 1987], which cautions against seeing the price, including tax, of the cheapest car as the key to the level of demand. Even after the tax reduction the number of cars in use per 1000 population in Germany reached only 19, compared with 42 in Britain and 41 in France. Although the Second World War was to postpone the realization of wide car ownership in Germany, the key in the long term was the 'people's car'. This was the outcome of a fertile collaboration between the accomplished car designer Ferdinand Porsche and Adolf Hitler, who ordered the Labour Front to arrange for the design of a low-priced family car for mass production. Although Porsche, to whom it fell to achieve this objective, had long been working towards a similar end, the impetus provided by Hitler intensified the focus, hastened the process, and ensured the funding to ensure a rapid and successful outcome. This was the origin of the post-war Volkswagen 'Beetle' [Overy, 1973; Blaich, 1987].

Historians' assessments of the industry's interwar performance range from failure to qualified success. Alford compared per capita car ownership figures in 1938 for Britain with the much higher ones of the United States, implying a marketing failure by the British industry [Alford, 1972, 1981]. Like others he dwelt especially on the backwardness of the industry by comparison with that of America. There, the sheer scale of standardized production enabled Ford, producing about half a million Ford V 8s in Detriot, to deliver in Europe at a price 30 per cent lower than the price of the same model manufactured at Dagenham, where fewer than 4000 were produced in a year [Rhys, 1972]. Other historians, while not disputing the facts of Anglo-American comparisons, have questioned the usefulness of choosing the United States as a comparator. They argue that the American industry had evolved within a market of such contrasting size, in terms of population, income, geographical distances and terrain, and where fuel was abundant at low cost, that superiority offers no surprise [Church, 1981; Bowden, 1991]. By virtually every criterion the British motor industry failed when compared with the United States. But by comparison with European competitors the interwar period saw manufacturers in Britain catch up and surpass French production in 1930, retaining an overall lead in Europe until 1956 (see Figure 1).

The disparity in car ownership between European countries also narrowed, though it was not until 1963 that Britain achieved the level of 7 residents per car which was the level attained by the United States as early as 1924. That figure was reached in France in 1962, in Germany in 1964 and in Italy in 1967. Market penetration was affected by differences in levels of private car taxation, import duties and credit terms and facilities. These meant that differences in running costs added to already existing intra-European disparities in real incomes and are fundamental in explaining international patterns of ownership density [Foreman-Peck, 1981; Bowden, 1991]. Within Europe average real wages were highest in Britain, where the terms and conditions of hire purchase available from the 1920s were also as favourable as

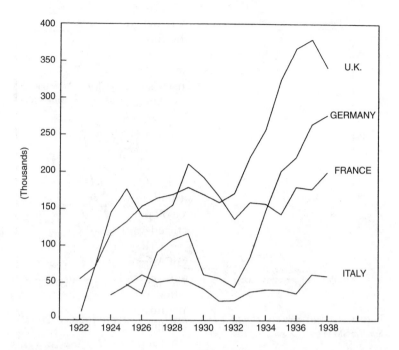

Figure 1 *Car production in Europe, 1922–38*
Source: SMMT

anywhere in Europe. Average car prices in Britain were marginally lower than in other European countries [Bowden, 1991].

After the slump Britain became the leading European exporter, accounting for about a half of all cars exported from Europe (including Britain) in the 1930s (see Figure 2). The markets of European car-producing countries were protected by tariffs, which were higher in France than in Britain throughout the period, as they were in the Fascist states in the late 1930s [Jones, 1981]. The British industry's 'failure', therefore, was of a similar character to that identifiable in other European countries: an inability or unwillingness on the part of manufacturers to penetrate beyond the upper and middle income groups in a context of overall limited market demand constrained by levels of real income [Church, 1981; Bowden, 1991].

The extent to which this represented a failure on the part of manufacturers is disputed. Historians who have emphasized the

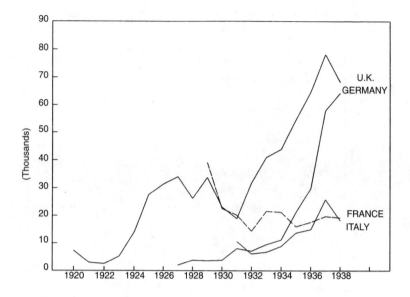

Figure 2   *Car exports, from the principal European countries, 1920–38*
Source: SMMT

industry's shortcomings on the supply side, especially those who
have focused on Anglo-American comparisons, are critical of British manufacturers for not having followed the Fordist approach to
achieve mass production for a mass market. Others who stress the
similarities between European industries, at least until the late
1930s, remain sceptical of the idea that a transformation of the
British market so as to resemble the market for cars in the United
States was a real possibility.

*(iv) Fordism and the British system of mass production: technology and
labour*

The most detailed exposition of the failure thesis does not, however, begin from such assumptions but instead draws heavily on the
labour process literature which stresses the centrality of the institutional context of production. In a major, valuable, comparative

20

study of the industry, and in a series of articles and essays, Lewchuck's interpretation of the dynamics of the British industry since its foundation has offered a fresh and challenging analysis. Its starting point is a model of technical change in which the 'effort bargain' between employer and worker is crucial. The outcome of interaction between employers and workers within a changing institutional context, the struggle between capital and labour is seen to determine which group controls effort norms, to what degree, and with what effects on capital investment [Lewchuck, 1987].

Even before 1914, Ford's sale of American vehicles assembled at Trafford Park achieved considerable success in tapping a hitherto latent market for low-priced vehicles designed to provide basic transport. Lewchuck argued that the consequence of the failure of British manufacturers to follow this strategy proved permanently debilitating and ultimately in the 1970s disastrous [Lewchuck, 1986, 1987]. Specifically, Lewchuck was critical of a failure on the part of British manufacturers to adopt a Fordist system of mass production which he considered to be the key which could have unlocked latent mass demand before 1939 as it had in the US.

That system consisted not only of mass production of standardized units but also involved a high degree of direct control over the labour force. This was necessary in order to realize the full potential for large-scale economies by achieving continuous, closely supervised flow production. Both at the Ford plant at Dearborn in the US and at Trafford Park trade unionists were sacked and other workers dissuaded from union membership by the offer of high day wages. Between 1912 and 1932, when the Dagenham plant opened, the Manchester factory employed only non-union labour [Wilkins and Hill, 1964]. Dagenham, too, was free from union influence until 1944, when the Labour Minister, Ernest Bevin, compelled that company, and others, to discontinue anti-union policies [Lewchuck, 1986].

Until that time, at least, Lewchuck presented the 'Fordist system' as one which gave maximum authority to the 'new managerial class'. Systematic, direct managerial control over labour involved strict supervision, which included the monitoring of workers by Ford security police, instant dismissal for trivial offences, unpaid layoffs, and the casualization of some workers. High day wages were systematically and unilaterally fixed by managers on the basis of time and motion studies [Friedman and Meredeen, 1980; Beynon, 1973; Thornhill, 1986]. Lewchuck contrasted these conditions with the 'British system of mass production', in which management relied on piecework to determine the pace of work, a system which gave workers the responsibility for setting production targets by linking earnings directly to output – and effort – expended on the job [Lewchuck, 1986, 1987].

According to Lewchuck's account, beginning in 1914 British manufacturers, most of whom were members of the Engineering Employers' Federation (EEF), attempted to create the conditions for a Ford-style industrial strategy. Shop stewards in Coventry, a major centre of engineering employment where the movement was increasing in strength, campaigned against a Fordist strategy of direct control and machine pacing, ultimately forcing employers to concede a degree of control over effort norms by falling back on piecework systems, leaving the decision governing production levels to individual workers. In Lewchuck's interpretation this basic structure of labour management, in which the balance of control lay with labour, was established during negotiations between the Engineering Employers' Federation and the trade unions between 1919 and 1922. These occurred during a period when the trade unions and shop floor committees enjoyed newly gained power resulting from the overheated labour market during the war, and during the immediate post-war boom.

The labour management system adopted by employers was also interpreted as a strategy by which British vehicle makers sought to create a sense of mutual interest between employers and workers [Lewchuck, 1987]. From that time, Lewchuck argued, managerial authority in British-owned car factories was less than complete, and the ability of British manufacturers to create an industry comparable in productivity and commercial penetration with their counterparts in the United States was damagingly crippled [Lewchuck, 1986, 1987].

This strong hypothesis has been criticized from several directions. Tolliday argued that when viewed in historical context piecework may be seen to have been as coercive as daywork, and that both direct control and piecework could operate as tight or slack systems, leaving considerable scope for variation in form and effect [Tolliday, 1987a, b]. British manufacturers' attempts to extend piecework represented a longstanding drive on the employers' part to gain control over the terms and conditions under which new techniques and practices should be introduced. The achievement of the piecework systems imposed at the time of the major engineering lockout of 1922 was perceived by employers as a victory, winning the right to manage through payment by results [Zeitlin, 1980]. A further question that needs to be pursued is whether, after the war ended, organized labour was sufficiently strong to impose and maintain a significant degree of control over work pace and productivity.

Following the national engineering stoppage in 1922 a period of feeble trade unionism allowed employers to apply semi-automatic and automatic machinery and to introduce flow production methods. Ford, which was not a member of the Engineering Employers' Federation, was blatantly anti-union, as were the two other major companies outside the EEF, Morris and Vauxhall, though both also practised a form of welfare to secure workers' compliance [Whiting, 1983; Holden, 1984; Thornhill, 1986]. As a member of the EEF, Austin was required to recognize the trade unions, but organization at Longbridge was discouraged and sometimes penalized [Church, 1979]. Craft workers, the most highly unionized section of the industry before 1921, were a dwindling proportion of all car workers between the wars, especially outside the body and machine shops. Clayden estimated that by the mid 1930s between 60 and 70 per cent of workers in the industry were semi-skilled, including women and boys; in the volume production factories of Austin and Morris the figure was between 70 and 80 per cent [Clayden, 1987].

Not only did the numerical strength of trade unions decline in spectacular fashion but newcomers entering the industry from outside engineering, and from rural districts around Oxford and Luton, were generally acknowledged to be willing to work hard in return for earnings almost twice the average levels of agricultural workers, and well above those in coalmining from which labour was also attracted [Chapman and Knight, 1953]. The minimum weekly wage for labourers on Oxford farms varied between 28s. and 36s., compared with 70s. and 80s. for Cowley car workers, whose hours were slightly less but their employment more irregular [Whiting, 1983]. During the 1930s the motor towns attracted huge numbers of migrants from the depressed areas, notably Wales and Scotland, but their militant trade union traditions do not appear to have had an immediate effect on trade unionism in the motor industry. Clayden concluded that the migrants' contribution was in providing organizational experience and leadership once local workers had taken an initiative; they did not play key roles in initiating action [Clayden, 1987].

Their opportunities, in any case, were limited, for unionization remained stunted in a period when semi-skilled and craft workers at Morris Motors were paid at rates some of which were around 30 per cent above the area levels for craft workers agreed between the Federation and the National Union of Vehicle Builders [Whiting, 1983; Thornhill, 1986]. High earnings diminished the incentive to organize, as did the threat of victimization by employers, who with few exceptions maintained an anti-union stance. Moreover, those unions which possessed the greatest potential for organizing workers in the motor trades, recruiting mainly from woodworking trades, skilled heavy engineering, docks and transport, accorded the recruitment of motor workers a low priority [Clayden, 1987].

Given the weakness of trade unionism between the wars, therefore, and in spite of its recovery in the late 1930s, Lewchuck's claim that 'British labour would not tolerate a managerial strategy which stripped them of any control over shop floor decisions' [Lewchuck, 1987, *183*], does not seem convincing. The weakness of the trade unions, and the virtual absence of shop stewards from most car factories between 1922 and 1934, not only removed labour's ability to improve conditions but enabled employers to ignore those elements of joint regulation embodied in the 1922 National Agreement between the EEF and the unions. Specifically, whereas mutual agreement between management and labour was required in fixing piece rates, which should be adjusted only when production methods altered, employers proceeded to impose piece rate changes unilaterally, regardless of circumstances other than market advantage [Thornhill, 1986].

How far does Lewchuck's characterization of management–labour relations and production methods at the major motor companies present an accurate picture? And is the 'British system of mass production' a valid explanatory concept? On close scrutiny, the implied contrast between the British and the American system is an oversimplification. The claim that after General Motors took over Vauxhall in 1925 management strategy assumed

Fordist characteristics has been disputed by the historian of Vauxhall [Holden, 1984]. Not only was the degree of managerial supervision of workers reduced but a bonus payments system was introduced. These formed the basis of management–labour relations at Vauxhall, which were also accompanied by a 'pragmatic welfarism', which included profit sharing, life insurance schemes, sports and social clubs, works outings and social events [Clayden, 1987].

Neither do the largest British manufacturers conform closely to the model. Austin was the one which adopted a piecework incentive system conforming most closely to the model, yet an alternative to Lewchuck's interpretation of its significance is possible. The introduction of bonus incentives payments, which Lewchuck regards as a concession on the employer's part to secure workers' cooperation [Lewchuck, 1987], did elicit increased effort from labour which did contribute to higher productivity. The 'concession' of self-regulation, however, was accompanied by the payment of semi-skilled and some craft workers below district rates, and below the widely accepted notoriously low basic rates for engineering labourers contained in the EEF agreement.

As a consequence, intensive effort became necessary, rather than optional, if piece rate workers were to ensure reasonable earnings. Neither this policy of hard driving through the manipulation of piece rates, nor that of displacing craftsmen by semi-skilled labour, a process which proceeded further at Longbridge than at most other car factories, justifies the terms 'concession', 'cooperation' and 'self-regulation' in describing Austin's labour management. Between 1924 and 1928 heavy investment in mechanization and flow line assembly led to a reduction in the workforce by one-third in a period of rapidly expanding output [Thornhill, 1986]. Managerial control between the wars was sustained by workers' compliance which was withdrawn temporarily when rate-cutting went too far. Hence the unusually large strikes, which involved mainly unskilled workers, in 1929, 1936 and 1938. These, together with the large stoppage at Ford in 1933, were triggered in response to rate or wage cuts and accounted for 82 per cent of all days lost through strikes in the industry [Turner, Clack and Roberts, 1967].

At the other end of the British spectrum, Lewchuck characterized the managerial approach at Morris factories as having been 'outside the mainstream of British management thought', where payment by results was combined with direct control and some machine pacing on Fordist lines. Yet at Cowley, as at Longbridge when mechanized lines for axle assembly and for the assembly of chassis were introduced in 1928, the line speeds were controlled by management [Thornhill, 1986; Engelbach, 1927–8].

The managers of British companies did differ from Ford in their use of piece rates and bonus payment, but the differences were those of degree. Furthermore, even if, as Lewchuck argued, the Morris factories were in some respects an exception to the British system of mass production, the model itself becomes more vulnerable to criticism. For throughout the interwar years those plants accounted for at least 40 per cent of the output from the four largest British-owned, volume car-making factories. In any case, within the context of the British market during the interwar years, the drastic decline in Ford's share, which only began to recover in the late 1930s, showed that Fordism was inappropriate. As an approach which was both a marketing strategy as well as a production philosophy it lacked that flexibility which enabled British manufacturers to re-establish their dominance.

(v) The dynamics and limitations of 'personal capitalism'

The failure to adopt a Fordist strategy, the obverse of the 'British system', is also seen by Lewchuck to have been in part a consequence of shareholders' preference for dividends over investment, or as Bowden presents it, 'a tendency to play it safe, for short-term profitability' [Lewchuck, 1985a, 1986, 1987; Bowden, 1991]. Chandler has incorporated this version of financial short-termism into his broader model of 'the dynamics of personal capitalism', which he believed to have been the key characteristic of much of British manufacturing industry in contrast with the United States [Chandler, 1990]. The motor industry is offered by Chandler as an exemplification of the baleful effects upon investment resulting from the dominance of family firms or by enterprises led by dominant personalities who inhabited owner-managerial structures, which typically distributed a high proportion of profit to the detriment of asset growth.

The financial dimension of the Chandler hypothesis rests almost entirely on Lewchuck's analysis of the profit and dividend policies of motor manufacturers, the basis for his explanation of differential rates of asset growth between British and American vehicle producers [Lewchuck, 1985a, 1986, 1987]. It is true that by the 1950s Ford's assets, which in 1929 were below those of either Morris or Austin, had outstripped their combined asset value. Any explanation of the disparate rate of investment which produced such an outcome, however, must take into account the comparable rates of asset growth recorded by Morris and Ford during the 1930s despite the differences in corporate characteristics. The different effects of war on British and American companies is another factor which cannot be disregarded. Similarly, Lewchuck's concern that the capital–labour ratio fell in the 1930s overlooks the complication resulting from rearmament, which beginning in 1936 saw the transfer of motor workers to aircraft production. For this reason the Austin ratios which he used to demonstrate falling labour–capital ratios took no account of several hundred aero workers at Longbridge by 1938 [Thornhill, 1986].

Lewchuck acknowledged that the lack of data does not permit direct comparisons between the financing of British and American firms before 1919. None the less his conclusions have been incorporated in Chandler's robust interpretation of the dynamics of British industrial decline, which further justifies careful consideration of the foundations on which the model was based. Lewchuck argued that whereas British companies enjoyed access to a capital market attuned to, and enthusiastic about, public issues, American firms had to rely on private sources and retained earnings to finance investment. But access to public funds, accompanied before 1914 by risks of fraudulent promotion and financial fluctuation, was not critical to corporate growth. Neither Morris Motors nor Singer, producing nearly 60 per cent of cars in Britain in 1929, sought capital from the Stock Market before the mid 1930s [Saul, 1962; Harrison, 1981; Thoms and Donnelly, 1985; Overy, 1976].

On the basis of a sample of British motor companies' balance sheets for the period between 1919 and 1932 Lewchuck found a relatively low proportion of retained profits [Lewchuck, 1985a]. The sample, however, excluded private firms, which unlike public companies were not obliged to submit financial returns to the Registrar of Companies. For that reason Morris Motors was excluded from Lewchuck's list of pre-1926 firms. This is an important omission from any analysis seeking to generalize about the dynamics of the British motor industry, because Morris Motors, which was owned entirely by W. R. Morris (later Lord Nuffield) until 1936, was retaining an average of 80 per cent of pre-tax profits in the 1920s, partly reinvested but also used to accumulate large reserves [Overy, 1976].

For the period between 1927 and 1951 Lewchuck noted that Morris retained only 26 per cent of earnings generated [Lewchuck, 1987]. Closer scrutiny of the accounts reveals that a break in trend occurred from 1936, which coincided with Morris's first public flotation [Andrews and Brunner, 1955], and the decline of his personal holding to 18.8 per cent. This review of the profitability of Nuffield's businesses does not reveal long-term corporate behaviour which conformed to a particular structure of ownership and managerial control on the Chandler–Lewchuck model. In the context of that model the Nuffield deviance is important because the company was so central to the industry. During the 1920s Morris Motors produced some 38 per cent of all British cars; however, the sports model built by Morris's other car company, MG, added to the larger, more expensive cars produced by Wolseley Motors, acquired by Morris in 1926, increased the percentage to above 40. In 1929 the figure was 51 per cent of cars produced by the 'Big Six', at a time when the industry was virtually a duopoly [Maxcy and Silberston, 1959]. Since Britain's largest motor company does not fit easily into the Chandler–Lewchuck model of corporate behaviour, any explanation of the industry's performance which relies on that model invites scepticism.

Reservations also need to be made concerning the financial history of the Austin Motor Company, the second largest British vehicle manufacturer from the late twenties, which emerged from the war crippled by debt and from 1920 heavily geared. Until 1928, therefore, the company paid no dividends on preference shares and none on ordinaries until 1929 [Church, 1979]. In contrast with Morris Motors this was a period of immense financial difficulty for Austin which had threatened to bankrupt the firm in 1920/1. Yet it was during the 1920s that the company eventually paid off large debenture loans and invested heavily in re-equipment and reorganization for volume production of a successful small low-priced popular car, the 'baby' Austin Seven. In the 1930s Austin was

Table 2   *Proportion of Net Earnings Retained by 'Big Six' Car Manufacturers (%)*

	1929–33	1934–8	1947–56	1952–6
Morris	50	25	39	} 68
Austin	33	31	72	
BMC				
Standard	80	40	52	
Rootes			79	
Ford	20	23	79	
Vauxhall	72	42	74	

Source: Maxcy and Silberston (1959), Table 7 and Table 18.

retaining a lower proportion of profits than in the 1920s, but the company's financial difficulties and heavy investment programmes had left shareholders particularly hungry for dividends during the 1930s [Church, 1979].

A comparison between the proportion of earnings retained by British companies and the competing American subsidiaries reveals a more complex pattern than one consistent with the Lewchuck–Chandler hypothesis. At the very least the 'pattern' is ambiguous (see Table 2). Lewchuck interpreted Rover's high-dividend policy in the 1950s as a measure intended to avoid a possible takeover [Lewchuck, 1986]. This cannot, however, explain low profit retention throughout the industry in previous decades, for it was not until after the Companies Act of 1948, which required greater disclosure of companies' finances, that predatory bids began to pose threats of unwanted advances from corporate raiders [Hannah, 1983].

How valid is the distinction drawn by Chandler between, on the one hand, those enterprises managed personally or by families which typically resisted the adoption of a modern corporate structure in the pursuit of economies of scale and scope, and on the other those organizations administered by salaried, career managers? The distinction is central to Chandler's model of the dynamics of capitalism, for he argued that whereas in Britain managerial firms were more likely to adopt growth of assets as a major goal, personally managed or family-dominated firms favoured short-termism in the form of 'a steady flow of cash to owners who were also managers' [Chandler, 1990, *390*].

The largest, Morris Motors, was the company most completely under financial and ultimately managerial control of its founding entrepreneur who even after it became a public company in 1936 continued to be the major single shareholder and chairman until the BMC merger in 1952 [Andrews and Brunner, 1955; Sargant Florence, 1961]. The ownership of the Austin Motor Company, too, was dominated by its founder until the 1930s. However, his 22.4 per cent holding was dispersed after Lord Austin's death in 1941 so that by 1951 no single shareholder possessed more than 1.4 per cent of the voting equity. The 20 largest shareholdings (which included directors) dropped from 12.4 to 0.1 per cent [Sargant Florence, 1961].

Throughout the entire life of the Rootes Group the company continued to conform to the founder-owning family enterprise dominated by the two Rootes brothers. Even in 1951, and indeed long after, the largest 20 shareholders, but mainly the Rootes brothers, held 55 per cent of all voting shares, of which 34.4 per cent belonged to the chairman and managing director William (Billy) Rootes [Sargant Florence, 1961].

The Standard Motor Company, the other major British motor manufacturer in the 1930s, was also led by a manager who possessed a substantial financial interest in the company. John Black was trained as a solicitor but it was his experience in the army, followed by management of the Hillman Motor Company, which brought him success as the joint managing director of that motor company. After Hillman was taken over by Rootes, Captain Black had joined Standard as the assistant of its founder, Reginald Maudsley, whom he succeeded as managing director after Maudsley's death in 1934 [Thoms and Donnelly, 1985].

In each case the management and control of the major British companies appears to have rested with the founder-owners who have been described as firmly in control of their firms until well after the Second World War [Hannah, 1976; Chandler, 1990]. This is undoubtedly true of Rootes until the company ceded control to Chrysler in the 1960s, and of Standard until Maudsley's death in 1934. However, in the two much larger companies the respective managerial roles of Morris (Lord Nuffield) and Sir Herbert Austin had diminished well before the war. At Austin the change began in 1921, when to avoid bankruptcy Sir Herbert reluctantly consented to share managerial power with two nominee directors put in place by the Creditors' Committee on the advice of the Midland Bank, one an engineering expert in production organization, the other a financial director. This left the founder to concentrate his attention on design [Church, 1979]. Whereas Morris Motors conformed to the stereotype of a 'personally managed' firm until the 1930s, share ownership in Austin was more widely dispersed. In 1925 this led to Sir Herbert lobbying shareholders in opposition to his co-directors who objected to his proposal to merge the company with General Motors [Church, 1979].

W. R. Morris's autocratic managerial rule began to weaken from the late 1920s. In recognition of the company's increasing dependence on the expertise of managers, in the 1930s he helped to reorganize the company in the form of a divisional structure which combined stricter central control with a more rational system of decentralized management of the subsidiary companies. His biographer noted that Nuffield 'accepted the usurpation of his throne for the sake of the survival of the enterprise' – though this did not mark the end of his personal interventions. These were erratic and sometimes perverse, affecting the company's strategy and its managerial personnel until the merger with Austin in 1952 [Overy, 1976]. The company's history from the 1930s raises doubts concerning the degree to which the multidivisional reforms of those years, or the 1947 organization chart which set down the new relationship between departmental heads, boards of directors and chairman, came into effective operation [Overy, 1976; Pagnamenta and Overy, 1984].

Although decision-making was increasingly left to boards of directors, comprised of a handful of managers whose equity ownership was minimal, founder-owners influenced business policies in the later years, despite organizational changes which at least implied limitations on their formal powers. Arguably equally important, is their continued influence on corporate strategies after their deaths through the senior figures whom they had appointed and whose roles during the critical post-war period were to be central to the industry's performance (pp. 84–92).

(vi) Fordism and the British approach to markets and marketing

One of the assumptions underlying Lewchuck's comparisons between Fordism and the British approach to mass production is that the large-scale, single, low-priced model strategy, had it been adopted by British manufacturers, would have created a market large enough to enable producers to achieve the economies of scale needed to justify investment in mass production [Lewchuck, 1986, 1987]. Supply factors, in other words, are deemed to have been paramount.

Other historians have attached particular importance to the differences between the market in Britain and that in the United States. Demand in the US was sufficiently large to enable American producers to secure economies of scale at each stage in the manufacture and assembly of parts and components. This was the case at Ford after 1910 and at General Motors from the late 1920s, even though that company's strategy was to produce a range of models aimed at different socio-economic groups [Rae, 1959]. Maxcy and Silberston, and Miller and Church have shown that throughout the 1920s the British market lacked social depth, a market limitation which was not overcome until the economic recovery lifted middle-class incomes, beginning in the 1930s. The extent of car ownership in Britain continued to lag far behind American levels and was overtaken by France in the late 1920s. At around 40 cars in use per 1000 people, ten years later ownership in the two countries was roughly similar [Blaich, 1987].

Whereas in the early 1920s price competition led by Morris was the key to a rapid growth in sales, during the 1930s design and model differentiation within price and horsepower bands was the principal determinant of companies' share of the market [Overy, 1976]. Even at relatively low levels of car ownership, by the late 1920s replacement sales, rather than sales to first time buyers, had become the major constituent of demand in what had become a 'mature' market. In order to deal with this development an important shift took place in British companies' market strategies, which began to pay increasing attention to the visual appeal of motor cars.

An increasing emphasis upon non-price competition was a response not only to a stagnating market but to technical innovation. Beginning in the late 1920s, the widespread adoption of closed saloon cars with roofs made of stamped steel made appreciable styling change possible [Maxcy and Silberston, 1959; Church and Miller, 1977]. Price stability in the short run replaced cut-throat competition, while design and styling, influenced to a degree by streamlining, which became popular in the US, became the weapons of business rivalry. Slight modifications to body contour (the origins of streamlining in Britain), accessories and colour were intended to yield an increasing number of visibly different variants for each of the basic standardized engines and chassis [Church and Mullen, 1989].

Even Austin cars, renowned for their founder's philosophy of utilitarian design (the industry was not, he told Austin agents at the launch of the Seven in 1922, 'a fancy trade'), succumbed to the model price competition practised by Nuffield. From 1929 Austin employed an Italian designer who undertook a radical redesign of the Austin Seven saloon car, the 'Ruby'. Conceived by Sir Herbert in 1921 as the ultimate utility vehicle, the 'Ruby' was described by its original creator as a necessary but regrettable capitulation to the vagaries of public taste – especially that of women [Church and Mullen, 1989]. The 1930s also saw a burgeoning of advertising, which included copy in magazines and newspapers, catchphrases and songs which drew attention less to mechanical reliability, which had been a preoccupation of the early 1920s, than to unique qualities and services. The purpose of the production of a full range of models and a profusion of variants was both to persuade buyers to retain brand (marque) loyalty among existing owners and to persuade them to switch to different makes of car. Such a policy was favoured by dealers in a market which was relatively stagnant and in which such a policy afforded greater protection to dealers' margins.

For a time, each of the major firms displayed at least one new model, or at least a 'facelift' at the annual Motor Show, a process which was costly for the volume car producers but presented even greater financial problems for the rest [Overy, 1976]. The relatively high cost of closed bodies, which incorporated styling changes, was due not only to additional capital cost but also to the added work by hand required to form the compound roof curves, and to the labour-intensive assembly and welding of the body and fitting the doors and windows. In the US the increase in the rate of model obsolescence which was a consequence of the annual styling introduced during the mid 1920s, had brought about a sharp decline in the number of small producers. This forced small manufacturers out of the industry, and eventually compelled Ford to abandon his single Model T marketing strategy [Thomas, 1973]. In Britain, however, the period of annual model changes was brief, ended by Morris in 1935. Thereafter the entire range of Morris cars, known as a 'series', were to remain unchanged, except for minor technical modifications, over several years until necessity required otherwise [Overy, 1976]. In the United States there was a continued reduction in the number of competing firms largely as a result of the cost pressures imposed by the annual model change. In Britain a few small producers secured a growing market share, their successful design and marketing strategies fragmenting the market through product differentiation [Church and Miller, 1977; Church, 1993].

During the growth phase of the 1920s, when the Morris Cowley, followed by the 'baby' Austin Seven, pushed down 'the pyramid of consumption', Morris and Austin came to dominate the industry. Whereas in 1922 their combined output was an estimated 13 per cent of all cars produced in Britain, by 1929 the figure was 60 per cent. Meanwhile, Ford's share of car production fell from almost 30 to 10 per cent. Clearly, Fordism in the form practised at Trafford Park, directed from Detriot, was a failure and continued to be so until well after the commencement of production at the Dagenham factory in 1932 [Wilkins and Hill, 1964].

Why did Fordism fail in Britain and why did the British system of mass production and its variants succeed at Ford's expense? Ford's single-model policy in the 1920s was based on a car designed for the American market. The low-cost Model T and its successors until 1932 were designed regardless of the fiscal differences existing in the two countries. In Britain the horsepower tax had the effect of raising the price of a Ford to a level closer to that of the new models produced by British volume manufacturers. However, throughout the 1920s, and indeed after the Model T's replacement from 1928 until the war, Ford's lowest priced model was the cheapest which sold in appreciable numbers. Until 1932 the approach of most British manufacturers differed from the Fordist market strategy by taking into account British consumers' preferences regarding performance, roadholding, running costs, appearance, and a basic degree of comfort [Church, 1981]. In the production of these models British manufacturers selected those elements of Fordism which seemed useful for flexible production policies. In 1925 C. R. F. Engelbach, Austin's head of production engineering at Longbridge, explained that company's manufacturing strategy in relation to the British market:

> A change has come over the spirit of our dreams of quick time floor to floor production performances, accompanied by the spectacular removal at miraculous speeds of chunks of metal to the musical ticking of stop watches . . . Rapid changes in fashion and ideas have slowed up the progress of special single operation machines. Continuous high production is too uncertain for special machines to be further developed. Designs have to be changeable at short notices . . . [and] at present there is probably no market likely to develop sufficiently that will lead to the extension of such specialized tool methods. [Engelbach, 1933–4, 7]

For this reason British manufacturers chose neither to invest in completely automatic transfer machines, which required long production runs to cover capital costs, nor to adopt machine pacing, which likewise depended on large standardized throughputs in order to be economic. Both Morris and, to a lesser extent, Austin opted for the flexibility offered by a low level of integration, purchasing a high proportion of parts and components from outside suppliers for assembly and finishing at Cowley and Longbridge. At the same time, Tayloristic subdivision of the timing and measurement of jobs and the production of standardized engines and chassis were features of volume production in British factories during the 1920s and 1930s [Tolliday, 1987b].

Whereas Lewchuck emphasized labour resistance and the 'underdevelopment of the managerial function' [Lewchuck, 1986] to explain the development and persistence of the British system of mass production, others have attempted to explain Anglo-American differences by stressing the character of the market and managers' approach to it. Kahn's view was that the deployment of resources in ways which led to the proliferation of models represented a failure of the industry's marketing strategy which was not conducive to growth in the scale of production [Khan, 1946]. This view was endorsed by Alford, who explained this trend in part as the result of the lack of personnel adequately trained to appraise the commercial value of technical advances regardless of their country of origin [Alford, 1972].

Bowden's econometric analysis of the interwar market for cars in Britain also dwelt on the adverse effects of British manufacturers' approach to the market. It prompted the conclusion that by failing to initiate the supply-side changes which might have enabled price reductions to take place, British manufacturers were limiting the size of their potential domestic market. Product differentiation in the form of model-price competition, producing several models each to compete with the models of other producers within a notional discrete price range, had the effect, she argued, of limiting the scope for economies of scale, lower costs and product prices which might have led to mass consumption [Bowden, 1991].

Others have argued that while this widely held view may be theoretically sound in terms of neo-classical economics, it ignores the reality of competition in the British market. In particular it overlooks Ford's post-war failure in Britain until the mid 1930s. In 1929 Ford accounted for only 4 per cent of car production in Britain, a percentage which did not rise appreciably until after the slump when new models manufactured at Dagenham and specially designed for the British market were introduced. The reduction in the price of the 8 h.p. Ford Popular by 25 per cent in 1935 reinforced this revival and a return to a 17 per cent share in car production in Britain in the same year. In 1938 Ford's 18 per cent share compared with 23 per cent for the Nuffield organization and 21 per cent for Austin (see Table 3). As percentages of car production of the Big Six, the figure for Ford was 19 per cent, for Nuffield 26 per cent and Austin 22 per cent [Maxcy and Silberston, 1959].

By that time Ford had increased the number of basic models and engines on offer to 4, which compared with 8 and 7 by Austin, and 17 and 10 by the Nuffield enterprises [Church and Miller, 1977]. Contrary to the counterfactual expectations which historians have linked to a Fordist single-model, low-price marketing strategy, no vast new market was created by the price cuts of the mid 1930s. Although Ford's share in production rose, the appearance of competitors' new models, beginning with the Morris Eight in 1934 and its successors, followed by the Austin Eight and Ten, the Standard Eight and the Vauxhall Ten, were accompanied by

Table 3 *Shares in Car Production in Britain 1919–38 (%)*

	1919	1921	1923	1925	1927	1929	1932	1935	1938
Morris/Nuffield	2	10	28	42	37	35	33	31	23
Austin		7	8	10	23	25	27	23	21
Ford		22	11	2	6	4	6	17	18
Rootes, Standard and Vauxhall						8	23	23	31

Source: Maxcy (1958), Table IV; Overy (1976), Table 1; Wyatt (1981), Table I; Church and Miller (1977), Table 9.2.

falling sales of cheaper Ford cars. They were outsold by more expensive models which offered greater comfort, performance, appearance and in some cases individuality. Sir Miles Thomas, commercial director of the Nuffield organization, remarked of his company's competitor to the £100 10 h.p. Ford Popular, the £100 Morris Minor, that 'no one wants to keep down with the Joneses' [Thomas, 1964, *168*], a reference to buyers' preferences for cars priced above that of the basic model in any given range.

That British, rather than American Fordist, marketing strategies were optimal under 1930s conditions is further suggested by the experience of those other European manufacturers who did imitate Ford. Both Berliet and Citroën in France and Fiat in Italy, 'dazzled by the Ford spectacle built capital-intensive plants designed for mass production of single models' [Tolliday, 1987b, *33*; Bardou *et al.*, 1982]. They underestimated the persistent consumer demand for quality and differentiation which also characterized the British market between the wars, causing serious financial problems for the Fordist imitators in Europe, bringing them to the brink of ruin [Tolliday, 1991].

If the production strategies may be understood as rational, for the British companies achieved substantially higher profitability than Ford, was there nonetheless a failure of innovation and marketing as has been suggested? An analysis of engine and model types, advertising copy and production figures, model by model, indicates that, with the exception of the Nuffield organization, the leading companies manufactured a much narrower range of cars

37

than has been supposed, by Khan for example [Khan, 1946; Church and Miller, 1977].

Many of the additional 'models' were made by the simple expedient of combining basic components in various ways. Nuffield was the major culprit, producing more than twenty models in the early 1930s, compared with seven from Austin and four from Ford. The contention that the industry failed to innovate, though, has been disputed. The American industry was the source of invention and design developments, but while such improvements as synchro-mesh gearbox, independent front suspension and unitary construction were first introduced widely in the production of Vauxhall cars in Britain they soon became available from other manufacturers [Church and Miller, 1977].

Sales methods, too, were influenced by American practices adapted to suit British dealers and car buyers. Even in the mid 1920s the larger British manufacturers were employing 'modern and intensive selling methods' of a kind which impressed a representative of the United States Bureau of Domestic and Foreign Commerce. These included hire purchase, available since 1920, extensive distribution networks, after-sales service, induction of sales personnel, and advertising, including company journals such as *The Morris Owner* and *The Austin Advocate* [Overy, 1976; Church, 1981]. Such evidence on marketing and business strategy practised by firms producing about one-half of total output has been offered as support for the view that within the context of the British (and European) market before 1940 the British manufacturers were, on the whole, rational in their responses, and relatively successful financially. Within the protected British market the slow growth of new-owner sales between the mid 1920s and the mid 1930s encouraged product differentiation as the favoured competitive strategy.

Ford's rapid progress from 1932, when cars produced at Dagenham were for the first time designed specifically for the British market, might suggest that Fordism was at last transforming the market. However, the simultaneous increase in the market share of Standard and Rootes and the slow growth in the 1930s of the lowest priced, lowest powered cars, point to a more complicated segmentation of market structure. Seen from the standpoint of competing car producers, within any given range price was not the single most important factor influencing consumers.

(vii) Debilitating environment: structures and strategies

One effect of the non-price competition of the 1930s, which was to have longer term consequences for the restructuring and rationalization of the industry, was that it enabled medium-sized firms, notably Rootes, Standard and Vauxhall, General Motors' subsidiary company, not only to compete but to increase their share of production. From 8 per cent in 1929 these three companies increased their share to 31 per cent by 1938 (see Table 3). The success of these firms depended in part upon the heavy use of bought-out supplies of components parts, mainly from large specialist producers, thereby giving assemblers access to external economies of scale. A report on the industry in 1950 gave examples of companies in which bought-out components were responsible for between 63 and 74 per cent of production costs, including raw materials [PEP 1950]. Variations in the degree of integration between major producers were important, but the PEP report made it clear that external economies were considerable by the mid 1930s. At the same time, by lowering fixed capital entry requirements the prevailing industrial organization helped to perpetuate the survival of the smaller manufacturers.

Foreman-Peck has drawn attention to another source of market failure, which enabled firms displaying competitive weakness, either through managerial failure or cyclical financial difficulty, to avoid the penalty of 'exit' from the industry. Adopting a similar approach to that of the Midland Bank towards the Austin Motor Company in 1920–1 [Church, 1979], in 1931–2 Lloyds Bank together with major suppliers, which included Pressed Steel and Lucas, agreed to continue financial support for the Rover Motor Company, an ailing, long-established specialist car manufacturer. The condition for rescue was the appointment of an independent accountant, subsequently elected to the Board of Directors, to investigate the company's affairs and make recommendations for improvements [Foreman-Peck, 1981a, b]. At the Standard Motor Company, too, 'voice' – a say in corporate policy in return for financial support – was the condition laid down by Barclay's Bank for their support in difficult times. The bank insisted on appointing two nominees to the board [Richardson, 1972]. Either out of loyalty or for fear of further losses in the event of bankruptcy, bankers contributed

Table 4   *Rates of Return on Capital by the 'Big Six' Car Manufacturers 1929–38 (%)*

	1929	1930	1931	1932	1933	1934	1935	1936	1937	1938
Morris	16	17	11	12	6	8	15	19	16	12
Austin	21	25	9	14	15	18	16	15	16	11
Standard	loss	19	24	48	27	19	25	23	26	9
Humber	loss	4	loss	loss	7	22	14	14	14	8
Ford	12	11	3	loss	loss	7	4	6	5	3
Vauxhall	n/a	n/a	12	18	41	54	58	47	32	24

Source: Maxcy and Silberston (1959), Table 6.

to companies' survival, thereby perpetuating an industrial structure which saw a decrease in the level of concentration during the 1930s.

Technical internal economies of scale, therefore, did not guarantee market dominance. What they could do, as the Dagenham plant showed from the mid 1930s, was to provide low-cost production. But Ford's sales and profit experience demonstrated that such an advantage only ensured survival with unstable sales and low profitability [Maxcy and Silberston, 1959]. Between 1929 and 1938 Ford's rate of return was consistently far below that of the other Big Six vehicle manufacturers (see Table 4). The flexibility of organization in the vertically-disintegrated structure of the volume car industry enabled car assemblers to purchase supplies from parts and components manufacturers operating on a larger scale. This made the specific performance of the main competitors more a function of individual marketing strategies and management abilities than scale economies of car production.

The expansion of the industry was chiefly the result of the internal growth of firms. Morris had purchased the bankrupt Wolseley company in 1926 and acquired the little specialist car-making firm, Riley, in 1938, but only Rootes among the Big Six car-making firms between the wars was the result of mergers of several existing companies – Hillman, Humber and Commer Cars. The commercial success of Rootes, Standard and Vauxhall during the 1930s resulted in an industrial structure which under different market conditions after the Second World War would make rationalization problematical. In 1938 the big three companies, Nuffield, Austin and Ford, accounted for 62 per cent of the output of cars in Britain. This compared with 69 per cent by the largest three producers in Germany and 73 per cent in France. Once again, however, the difference between the much higher ratio of industrial concentration in the United States, where the figure was 85 per cent, is significant in underlining the greater similarities between the major car manufacturing countries in Europe before 1939 [PEP, 1950]. Nonetheless, a major focus for contemporary commentators, and subsequently for historians reflecting upon the industry's prospects and performance after 1945, was the debilitating effect of the industry's structure.

On the eve of the First World War, the British motor industry, then organized in a large though diminishing number of competing firms, was under threat from the Model T Ford, assembled from kits imported from the US. Ford's output in Britain was twice that of the largest British producer while the American company took a rapidly rising share of the home market. A handful of British manufacturers had begun to make smaller, lower priced cars in quantity to compete, but to an even greater extent compared with other European producers the scale of production of the largest enterprises was counted in thousands rather than tens – and for Ford hundreds – of thousands. Under peaceful free trade conditions the output of the industry in Britain was rapidly catching up with that of France. However, the principal contributor to this process was Ford.

By the outbreak of the Second World War the industry had become the largest in Europe. Protection and, by European standards, buoyant middle-class incomes, particularly in the 1930s, had enabled the infant British industry to establish volume manufacture comparable in scale to the largest manufacturers in Europe. Critics pointed to the continuing contrasts with the much larger size of plants and firms in the United States. There the higher degree of standardization and the concentration of production on fewer models had enabled American manufacturers to achieve levels of efficiency far above those in Britain, where cars were still only affordable by a limited middle class. The National Advisory Council (NAC) initiated in Whitehall made all of these points in its 1945 Report, although it concluded that: 'The industry has shown by its past performance that it is in itself vigorous and efficient . . . and if its future environment is such as to apply the appropriate stimuli, there is every reason to expect that it will respond to them vigorously and effectively' [NAC, 1947, 9]. This more optimistic note no doubt owed something to the presence on the NAC of an eight-member majority representing the Society of Motor Manufacturers and Traders.

2   The Roots of Decline

(i)   Post-war pre-eminence: attainment and erosion

The immediate effect of war was a reduction in the output of cars to very low levels in order to facilitate the production of goods-carrying vehicles to meet military requirements. The Shadow Factory Scheme, financed and introduced by government in 1936, had already provided extra plant and equipment for aircraft manufacture by motor firms, to which was added the production of tracked carriers, tanks, tractors, and a wide range of military goods. New buildings and much of the machinery installed for these purposes expanded the industry's motor manufacturing capacity after the war [Maxcy and Silberston, 1959; Thoms and Donnelly, 1985]. The strength of the trade unions and in particular that of the shop stewards' movement, was one wartime development which was to have longer lasting effects on the industry than it had following the First World War. These developments were powerfully reinforced by the 1939 Emergency Powers Act, which made it more difficult for employers to dismiss labour, and by the role given to shop stewards by the Joint Production Committees which employers were obliged to set up [Flanders, 1952]. Another development of long-term and industry-wide importance was the 1941 Coventry Tool Room Agreement, reached between the Amalgamated Engineering Union and Coventry employers at a time when fierce competition for labour caused the pay of semi-skilled production workers and dilutees to exceed that of apprentice-trained toolroom craftsmen. The accord, which effectively protected skilled workers' differential rates relative to those paid to production workers, remained in operation until unilateral termination of this agreement by the Coventry and District Engineering Association in the early 1970s and was to have major repercussions throughout the industry (Thoms and Donnelly, 1985).

Home demand after the war was buoyant but constrained by government controls on the supply of steel which was channelled to firms which exported 50 per cent (raised in 1947 to 75 per cent) of their output. This policy was part of the government's strategy to counter the loss of dollar earnings resulting from the

43

war by a systematic export drive. A sellers' market was created by the absence of competition from Europe, where productive capacity was slower to recover from the effects of defeat or occupation. This was combined with a resurgence of pent-up demand in the United States and some Commonwealth countries, notably Australia, which American producers could not satisfy. The result was a surge of British vehicle exports at unprecedented levels to which devaluation in 1949 brought a further boost [Maxcy and Silberston, 1959].

For a time Britain was the world's leading car exporter. The industry's share of world motor exports rose from 15 per cent in 1937 to 52 per cent in 1950. In that year 75 per cent of all cars and more than 60 per cent of commercial vehicles (mostly chassis and vans) were sold abroad. Yet car production was one-third greater and commercial vehicles almost twice as large in 1950 compared with 1937. This exceptional trading record lasted until the mid 1950s. By that time the American motor industry had resumed its virtual monopoly of the US market and European manufacturers had also recovered. Germany overtook French car production in 1953 and that of Britain in 1956, when Germany's long-term dominance in Europe commenced (see Figure 3).

Thereafter, Britain's share of world trade in cars averaged 24 per cent between 1957 and 1962, falling to 19 per cent in 1963–7. The comparable figures for trade in CVs were 28 and 30 per cent. British production as a proportion of output from the major manufacturing countries fell from 11.4 in 1960 to 8.5 per cent in 1970. The positive balance of trade in motor vehicles, which continued, though declining, until 1977 was increasingly attributable to buoyant commercial vehicle exports. From 1974, for the first time since 1914, the value of car imports exceeded exports [SMMT]. These developments occurred within a context of the recovery of European production capacity and a progressive, international reduction in protection.

Market development also played a part. In 1938 the number of cars in use per thousand population was 19 in Germany, compared with 41 in France and 42 in Britain [Blaich, 1987]. By 1960 the figure for West Germany was 84, compared with 119 in France and 106 in Britain, although during the 1960s car density in West Germany overtook that in both countries [SMMT]. By 1970 the real GDP per capita of Germany and France was respectively 75

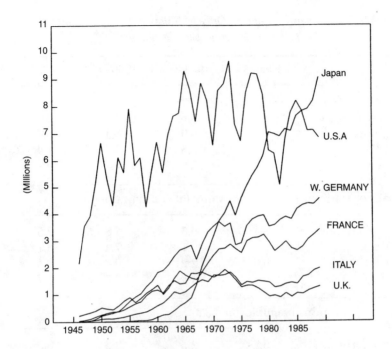

Figure 3 *Car output in the principal producing countries, 1946–89*
Source: SMMT

and 72, compared with 63 in the UK (as a percentage of US purchasing power parity), the outcome of higher rates of growth in
those two economies in each quinquennium throughout the
1960s which continued in the 1970s [Jones, 1981] (see Table 5).
The slackening in the rate of increase in car ownership in Britain
was reflected in the falling proportion of sales to new owners,
which dropped from some 72 per cent in 1960–4 to 34 per cent in
1970–4, leaving replacement demand once again the main determinant of sales. In Britain the rate of growth of the total stock of
cars rose by more than 10 per cent between 1953 and 1963 before
falling to 8 per cent in 1963–8 and 4.5 per cent in 1968–73. This
compared with an annual average of 12 per cent growth in the
major European markets throughout the period [HC, 14th
Report, 1975].

45

Table 5 *Western European Growth Rates of GDP and the Motor Industry (annual average per cent change)*

	1960–4	1964–9	1969–73	1973–8
GDP				
W. Germany	5.1	4.6	4.5	2.0
France	6.0	5.9	6.1	3.0
UK	3.1	2.5	3.0	1.4
Italy	3.5	5.6	4.1	2.1
Motor car industry output by value (at constant prices)				
W. Germany	8.1	6.5	4.8	2.5
France	5.5[b]	9.1	7.4	3.1
UK	5.6	2.2	0.4	−2.4
Italy[a]	n.a.	n.a.	3.6[d]	1.7
Motor car industry employment				
W. Germany	6.2	2.3	2.3	0.5
France	2.5	2.6	5.1	0.7[e]
UK	2.4	0.9	1.3	−1.4
Italy	4.9[c]	6.8	1.5	1.4[e]
Motor car industry output by units produced				
W. Germany	9.1	4.4	2.3	1.2
France	4.2	8.8	7.0	1.7
UK	6.5	−1.3	−0.2	−4.7
Italy	14.0	7.9	5.2	−2.9

Notes:
[a] Relates to transport equipment, not just the motor vehicle industry.
[b] 1962–4
[c] 1961–4
[d] 1970–3
[e] 1973–7
Source: Jones (1981), Table 1.

To some extent, therefore, disparity in the rate of market development between Britain and continental competitors was merely a symptom of catching-up, made possible in part by trade creation within the EEC where incomes were growing more rapidly than in Britain. The differential stimulus to investment within the separate protected British and EEC markets was reinforced during most of the 1960s by an overvalued pound sterling. Between 1956 and 1968 the 33.3 per cent tariff on cars was reduced to 22 per cent, while under the Kennedy Round the general tariff fell to 11 per cent by 1972. Entry into the EEC in 1973 heralded the complete removal of tariffs between Britain and the major west European vehicle-producing countries in 1977.

By that time the decline in the British motor industry was well advanced, although the possibility of such a deterioration in its international competitiveness had been mooted shortly after the war ended. In 1945 a Whitehall committee had drawn attention to the industry's poor export record before the war, and commented on the division of the industry into 'too many, often small scale units, each producing too many models' [Barnett, 1986, *165*]. Two years later, the National Advisory Council for the motor industry predicted that when the sellers' market of the post-war years ended, higher productivity would be a necessary condition for continued success in export markets [NAC, 1947]. Both the Council and the report on the industry by the advisory body to the Ministry of Supply (PEP) drew attention to the fragmented structure, the excessive variety of models, lack of standardization of components, and the high unit costs associated with small-scale production relative to American car-makers, features which they argued must be changed if the industry was to compete internationally. This was, of course, a problem which had its origins in the particular form which non-price competition had taken during the 1930s.

Criticisms from Whitehall had prompted the SMMT to promote standardization of components and parts used in the industry, but it was left to individual manufacturers to implement agreements and they showed little interest. For a brief period the Nuffield and Austin companies participated in the scheme, but cooperation broke down when Lord Nuffield had second thoughts about pooling technical information which might prejudice his organization's competitive position [Overy, 1976; Jeremy, 1984–6]. Ten years later two leading economists drew similar conclusions to those of PEP concerning the structure of the industry, which they described as unsuited to the economics of car production for a mass market. They maintained that the industry consisted of too many firms producing too many different models even to approach the technical optimum level of production [Maxcy and Silberston, 1959].

Comparable, and possibly marginally higher, productivity was one of the factors which explains Germany's competitive lead in production from 1957. However, calculations of international productivity differences are fraught with difficulty. For the pre-war years the lack of adequate statistical data has forced historians to use crude productivity estimates, dividing the total number of vehicles produced by the numbers employed in the industry. The flaw in this approach lies in the disparity in the typical sizes of vehicles in different countries and in the raw material inputs they embody. Pratten and Silberston devised a weighting system to take vehicle size into account, measuring productivity in terms of 'vehicle equivalents' [Pratten and Silberston, 1967].

A refinement of this weighting formula by Jones and Prais resulted in the international comparisons presented in Table 6. The figures shown for 1955 imply productivity differences of a similar order to those estimated by Rostas for 1935/6 [Rostas, 1948] and suggest that until the late 1950s the motor industry in Britain appears to have been broadly in step with that in other European countries. Ten years later the productivity of American plants in the United States remained four to five times higher than in European factories, but more worrying for those concerned with the competitiveness of the British industry was the disparity between labour productivity in the German industry (6.4) and that in the British industry (5.8), revealing a gap which increased further in the 1970s. Productivity estimates based on net output values and employment showed German productivity in 1970 to have been two-thirds greater than that of the British industry [Jones and Prais, 1978].

In 1958, when the beginning of a shift towards German superiority was barely discernible, a detailed study by Maxcy and Silberston concluded that the British industry still enjoyed a cost competitiveness though that was attributed to a considerable degree to the efficiency of large-scale component suppliers. As in other observers' earlier reports, however, they concluded that

Table 6 *'Equivalent' Motor Vehicles Produced per Employees per Annum 1955-76*

	Britain	Germany	United States
1955	4.1	3.9	19.8
1965	5.8	6.4	25.0
1970	5.6	7.5	19.6
1973	5.8	7.7	25.0
1976	5.5	7.9	26.1

Source: Jones and Prais, (1978), Table 4.

unless the scale of production of individual models was appreciably increased motor manufacturing might not continue to be 'the most progressive of British industries' [Maxcy and Silberston, 1959, *198*]. They also drew attention to overmanning, as well as to weakness in vehicle design, quality, salesmanship and service, which they thought had already begun to affect exports.

These criticisms were repeated in more forceful terms during the mid 1970s in a number of major official inquiries into the ailing industry. Differing in most respects only in detail in their diagnoses, there was general agreement that the lack of international competitiveness resulted from a failure to secure economies of scale, due to a considerable extent to the continued existence of too many models, too many plants and too much capacity [HC, Ryder Report; HC, 14th Report; CPRS Report, 1975]. Low productivity and a lack of investment which left the industry with obsolescent plant were other defects, although in terms of future government policy towards the industry it was the condemnation of overmanning, restrictive work practices and poor industrial relations which was regarded by government to be the key to initial improvement (pp. 100–1).

The decline of the industry falls into three stages. The first began in the 1960s with a series of mergers between British companies and the acquisition of Rootes by Chrysler to form Chrysler UK in 1967. This stage culminated in 1968 in the merger of British Motor Holdings (BMH) with Leyland Motors to form the British Leyland Motor Corporation (BLMC, later BL). From that time, with the exception of highly specialist manufacturers – Rolls Royce, Aston Martin, Lotus and Morgan – BLMC/BL comprised the whole of the British-owned motor manufacturing industry. The second phase continued until 1975, when the industry came under state control, thus marking the beginning of the third phase of decline. Historians and other commentators have differed in the emphasis which they have placed on the variables contributing to the industry's difficulties. These include the role of government, industrial relations, industrial structure, investment and marketing policies, business strategies, the quality of management, and the role of multinationals.

(ii) Private investment and public policies: government and industry

The role of government is central in some accounts of the industry's decline. Successive governments' policies towards the motor industry after 1945 had the effect, Pollard argued, of corroding enterprise. 'Thirty years of discouragement had by the 1970s accustomed manufacturers to a low-growth, low-investment economy' [Pollard, 1982, *127*; see also Dunnett, 1980, and Barnett, 1986]. Those policies included the imposition and variation of purchase, and later value added, taxes, hire purchase restrictions affecting minimum deposit levels and periods for repayment, interest rates, incomes and regional policies [Dunnett, 1980]. Why was the industry seen to be central to the success of government macroeconomic policies designed, at one time or another, either to promote exports and improve the balance of payments, to sustain the exchange rate, to control inflation, or to protect employment?

The resilience of the industry during the interwar period continued to impress contemporaries for more than twenty years after the Second World War, and was part of Rostow's justification for describing the industry as a 'leading sector' in the economy until the 1960s [Rostow, 1963]. The value of net output rose from £300m to £639m between 1954 and 1966, an increase from 3.4 to 5.1 per cent of all industrial net output. This represented 7.5 per cent of all manufacturing production and 6 per cent of all manufacturing employment. A growth of 113 per cent during that period compared with an increase in industrial production of 41.5

per cent. It had been estimated that between 1954 and 1966 9 per cent of the economy's growth was attributable to motor vehicles, investment in the industry representing slightly less than 10 per cent in all manufacturing industry [Armstrong, 1967].

By the 1960s total direct employment in the industry was close to 0.5m, rising by 33 per cent between 1959 and 1973, mostly located in the West Midlands and in the South-East [Durcan, McCarthy and Redman, 1983]. Figures for total employment in the industry, including other supply and component sectors (though excluding employment in sales, repair and maintenance) brought the number to 0.8m in 1973 [HC, 14th Report, 1975]. Throughout the post-war period the motor industry grew faster than gross domestic product, as it was to do in France, Italy and above all in West Germany, until the early 1970s [Jones, 1981]. In 1963 one estimate based on input-output analysis of the combined contribution of indirect and direct inputs to total industrial production was nearly 11 per cent, a figure accepted as plausible by an official committee investigating the industry a dozen years later [Armstrong, 1967; CPRS, 1975]. Nearly one-third of industrial growth in the economy in the 1950s and 1960s has been attributed to the motor industry and its suppliers. The state of the motor industry, therefore, was a major influence through the multiplier effects of a wide range of interdependent industries [Armstrong, 1967].

In what ways, then, did government policies affect development and international competitiveness after 1945? Corelli Barnett's robust critique of government, aimed specifically at the Coalition and Labour governments of the immediate post-war years, focused on a search for a 'New Jerusalem' in the form of the welfare state which absorbed both attention and resources which should have been applied to industry. The motor industry, he argued, suffered particularly badly from a mixture of a lack of interest by the politicians or at the very least by 'tinkering' as a substitute for policy [Barnett, 1986]. One very important piece of tinkering was the setting of high export targets, which in 1948 rose to 75 per cent of each company's output, as part of the national drive to earn dollar currency. In the pursuit of this objective the government imposed penalties, in the form of restricted raw material supplies, rather than extending substantive support to British car manufacturers. Short-term instrumentalism of this kind Barnett interpreted as evidence of a complete disregard for the need for a national, long-term, industrial strategy, a part of which should have been rationalization.

To some extent this is to confuse outcome with intention. The National Advisory Council (NAC) was the creation of Whitehall in 1945 and produced a report which pointed to the need for the industry to move towards standardization, for systematic analysis of export markets and improved designs to meet consumers' requirements. The lack of progress in these directions, favoured by Whitehall, is explained by the resistance of the industry to intervention which involved any measure of compulsion. That position was successfully imposed on the NAC by the SMMT representatives, who from its inception outnumbered others on the Council. It remained, therefore, no more than a channel of communication [Tiratsoo, 1992].

Dunnett maintained that the government's post-war export drive contributed much to achieving the national objective of dollar earnings. Yet the method of allocating quotas of scarce steel froze the industry's fragmented structure and organization by, in effect, protecting the least efficient. Quotas were allocated to firms regardless of size or production costs, the sole criterion being the ability to meet annually-set export targets. In an immediate post-war sellers' market and under pressure to accelerate production for exports no new models appeared on the market until the popular 8 h.p. Morris Minor in 1948.

When the war ended, the pressure of demand was greatest in non-European markets to which British pre-war models were unsuited. Nonetheless, the immediate pent-up demand for cars in short supply worldwide, coupled with the necessity to reconstruct European production capacity, enabled British producers to re-enter imperial markets. It also enabled them to establish a foothold in the US, where for a time British models filled a niche as a second car for local use [Maxcy and Silberston, 1959].

British manufacturers had long argued that the horsepower tax penalized high-powered vehicles of the kind popular in the United States and the Empire, placing them at a disadvantage in export markets. Cars with high-speed engines, narrow track, low ground clearance and limited luggage space, which were suitable for the home market, were less suited to European markets. Yet the financial risk attached to developing cars specially designed for export deterred such a strategy [Church, 1979]. It seems more likely, however, that the profitability of the protected home market in the 1930s had provided insufficient inducement to manufacturers to develop overseas markets outside the Commonwealth. In any case, the outcome of Lord Nuffield's patriotic attempt to cultivate exports suggests a flawed perception of consumer preferences outside Europe. The 'Empire Oxford', which was conceived by him and built specifically for the overseas market, was returned unsold in large numbers and the project abandoned as a failure [Overy, 1976]. Within ten years, however, the Volkswagen Beetle, already selling widely in Europe, with no concessions to the conventional requirements of automobiles sold in the US had established a valuable and enduring foothold in that rapidly expanding market [Nelson, 1967].

In 1946 government accepted the arguments of the motor manufacturers by replacing the horsepower tax with a flat rate tax. Nonetheless, it was the popularity of small cars in the home market which increased, though that trend went into reverse during the 1960s. This suggests that the discriminatory horsepower tax was less critical than other determinants of consumer demand, particularly income levels and running costs, in influencing the structure of car sales at home and in overseas markets. Manufacturers' allegations (implicitly accepted in the 1948 tax reform) that exporting was hampered by the horsepower tax limiting sales of larger, more powerful cars are unsupported [Rhys, 1972].

During this period the reputation of British vehicles suffered from the shipment of unsuitable and defective models by manufacturers who were short of high-quality raw materials and under pressure to meet export targets. In 1950 British commitments to the Korean War led to a sharp increase in the demand for steel for rearmament. As before, quotas were reintroduced which, in common with orders for military vehicles, were placed irrespective of allocative efficiency. Again the industry's structure was reinforced by the same Ministry of Supply that had criticized fragmentation in 1947 [Dunnett, 1980; NAC, 1947].

The contrast with West Germany is revealing. The Volkswagen plant was built at Wolfsburg immediately before the war when it was planned to produce the 'people's car', or 'Beetle' as Porsche's low-priced utility car came to be known, at an annual rate of 150,000 units [Overy, 1975]. Between 1934 and 1939 Volkswagen benefited from Nazi policy which took various forms of discrimination against other firms and multinationals. The company came under the direct control of the Ministry of Munitions in 1943, a period which saw further diminution of workers' already minimal political and economic rights [Reich, 1990]. After the war, a particular British (Treasury-driven) concern among the occupying powers that the export capacity of German industry should be re-established rapidly to ease the Allies' burden of feeding and clothing the population has been held to have produced an unintended effect. This was to reinforce the relationships and institutions existing prior to the occupation, ensuring a continuing discrimination by the state in favour of Volkswagen. Contrary to the commonly-held view, while the quality of the Beetle occasioned scepticism among British observers they acknowledged that the Wolfsburg plant, virtually undamaged, was probably the most modern installation in the world [Reich, 1990].

In 1948 the German administrator appointed by the British to prepare the plant for peacetime production was succeeded by Heinz Nordhoff, formerly of General Motors' German Opel division. Under his management a new system of financial control and cost accounting, and a new distributor and dealer system were introduced, as was the practice whereby senior managers gave regular reports to workers on the company's overall progress and problems. Because of the limited home market and the Allies' stringent financial requirements Volkswagen's survival was virtually dependent on exports. From the post-war beginning, therefore, high-quality reliability and service, supported by an extensive distribution network across Europe, formed a major part of Volkswagen's corporate strategy [Nelson, 1967]. By 1956 two completely new large factories had been built, while a network of exclusive Volkswagen dealers on the Ford model offered service and parts. This network was well placed to exploit the fashion which developed among certain sections of the American population, notably the young, for inexpensive, small cars during the mid 1950s. Already by 1956 German cars sold overseas, almost entirely Volkswagens destined mainly within Europe but increasingly to the United States, exceeded Britain's total car exports [Nelson, 1967].

By comparison with British car manufacturers Ford emerged from the post-war period in a strong position. This was partly because of historical circumstances, in particular the construction of the new large plant at Dagenham, and partly because of Ford's access to the parent company's resources. When the new plant commenced production in 1932 its annual capacity was 200,000 units, a substantial proportion of which was planned to supply markets in Continental Europe, where, in accordance with Perry's long-held vision, the Ford subsidiaries were subordinate to the Ford Motor Company in England. The contrast between the scale and modernity of Dagenham and British car plants, most of which had developed from small factories built early in the century, was clear and remarked on in *The Times* in 1947, when Dagenham was described as a showpiece [Wilkins and Hill, 1964]. Not until the post-war boom did peacetime production reach full capacity and the economies associated with it, when the state-sponsored export drive enabled Ford to exploit its international distribution network to expand overseas sales.

In 1950 Ford was the leading vehicle exporter in Britain, a dollar-earning performance which government recognized by freeing dollars to enable that company to purchase American machinery and equipment unavailable elsewhere. The parent company's assistance also took the form of help, on a contract basis, in designing two new post-war models which entered the market in 1950 [Wilkins and Hill, 1964]. Novel in their symmetrical appearance and in construction, incorporating steel-welded bodies, independent front-wheel suspension and new hydraulic brakes, the Ford Consul, 20 per cent lower in price than other models of the same horsepower, and Zephyr set new standards [Adeney, 1989; Wood, 1988].

Whereas before the war both the levels and rates of growth in assets of Morris and Ford were similar, the immediate post-war years were a turning point. After 1945 investment in Ford saw a rise in net asset values which in 1956 exceeded those of the British Motor Corporation (BMC), formed in 1952 by a merger between the constituent parts of the Nuffield organization and Austin [Maxcy and Silberston, 1959]. BMC was still the largest car producer, accounting for some 39 per cent of all cars manufactured in Britain. But Ford's share was 27 per cent, having risen from 14.4 per cent (compared with Austin and Nuffield's combined proportion of 43.4 per cent) since 1946 [SMMT]. While the government may have given Ford special help to achieve relatively rapid preparation for post-war conditions,

Reich's view of this as an almost discriminatory treatment of the American MNE in comparison with dealings with British firms may be challenged on the basis of the immediate post-war history of the Standard Motor Company, the third largest British car manufacturer. Its forceful director's blueprint for an attack on overseas markets, based on mass production of a single model designed specifically for the export trade, met with substantial support from Whitehall. This took the form of dollars with which to buy machine tools, a low rental for a large, well-equipped shadow factory, and financial bridging arrangements to enable Standard to commence production of the Vanguard, launched in 1947. This suggests, contrary to Barnett's view that the government ignored the need for an industrial policy, that within the constraints of industrialists' implacable opposition to intervention at the industry level it was prepared, regardless of corporate nationality, to support companies which seemed likely to succeed. Standard's failure to achieve comparable success to that of Ford had more to do with internal corporate problems than to government [Tiratsoo, 1992].

Even though Ford was one of the two firms which it has been suggested received favoured treatment compared with others, the company was affected none the less by the same macroeconomic policies which several historians have identified as particularly deleterious, notably to investment and industrial relations. As the leading dollar-earner in the manufacturing sector, the motor industry was especially affected by general measures intended ultimately to protect the value of sterling by choking off home demand and restricting imports to improve the balance of payments. Hire-purchase terms, which since the mid 1920s were involved in most car sales and were an important determinant of demand, were changed seventeen times between 1952 and 1968. Adjustments in purchase tax were made thirteen times [Rhys, 1972; Dunnett, 1980].

Both contemporary and recent commentators have argued that these alterations were the cause of a demand instability which resulted in alternating periods of investment and expansion. One consequence was the underutilization of plant capacity which adversely affected productivity; another was the intermittent laying-off of workers in an industry in which employment for semi-skilled production workers had always been seasonal. The effect of the first was to increase manufacturers' unit costs, reduce profitability and check investment. The second contributed to a deterioration in industrial relations [Dunnett, 1980].

By maintaining the value of sterling government also increased the difficulty of exporting. Competition in third markets in which no indigenous motor industry existed against vehicles sold in undervalued currencies necessitated reductions in dealers' margins to be competitive. Consequently the attraction of a British manufacturer's franchise among dealers overseas was affected both by the limited turnover by comparison with the major European producers and by low profit-margins. The devaluation of sterling in 1967 followed by the revaluation of the Deutsche Mark in 1969 tended to remedy the imbalance, but by that time the large-scale production and extensive distribution systems of the leading European manufacturers, notably Volkswagen, were already well developed [Rhys, 1972].

Critics of government policy have argued that the uncertainty which frequent changes in fiscal and monetary policies had upon demand, growth, and profitability in the home market hampered investment and checked improvements in productivity [Bhaskar, 1979; Dunnett, 1980]. Estimates of productivity relative to the German industry between 1965 and 1970 show a ratio of German to British 'equivalent' motor vehicles rising from 1.1 to 1.3: the comparable American to British ratio was 3.5 (see Table 6 above). By that time the one remaining volume car manufacturer in British ownership possessed the lowest assets per man of any of its rivals in Europe and the next to lowest value added. The assets per man of Ford UK were the highest of any British car maker, but nonetheless ranked fifth after Ford (Germany), Opel and Volkswagen, and sixth in terms of value added [Rhys, 1972].

Both the Ryder Report and the Fourteenth Report of the Trade and Industry Expenditure Committee agreed that a substantial increase in investment was a necessary condition if the British motor industry was to match European levels of productivity. Comparing output per man with fixed assets per man, the Expenditure Committee found a significant correlation. However, using the same data, a statistical analysis of the relation between productivity and capital intensity in fourteen vehicle companies in various countries led Jones and Prais to conclude that the evidence on which the committee's view was based did not show that a greater capital input per unit of output was necessarily coupled with a higher output per unit of labour [Jones and Prais, 1978]. They insisted that the particular emphasis placed on capital expenditure as a cause of lower productivity in Britain was mistaken, and that capital utilization, which depended on organizational and human factors, was probably a more important explanation [Jones and Prais, 1978].

None the less, the difference in investment levels between major European companies was considerable. On an already different investment and technological base level, annual investment between the mid 1960s and the mid 1970s was £197.5m at Volkswagen and £117.5m at the state-owned Renault company, compared with £86m at BLMC. This discrepancy helps to explain the relative age of the British company's plant and equipment on which the Expenditure Committee commented and which, all things being equal, placed a ceiling on the productivity levels achievable. Productivity at BLMC was half that of Ford UK [Jones and Prais, 1978], a discrepancy which receives further consideration below.

To what extent can these international differences be explained by government policies affecting demand, as has been argued [Bhaskar, 1979; Dunnett, 1980; Pollard, 1982]? How far did taxation levels adversely affect home demand and production? After purchase tax was cut from 45 to 25 per cent in 1962 tax fluctuations were limited within the 20 to 33 1/3 range before 1970 [Dunnett, 1980]. Over roughly the same period these figures compare with between 11 and 21 per cent in Germany, between 13 and 28 per cent in Italy, and between 12 and 28 per cent in France [Rhys, 1972]. The differences, therefore, are those of degree, the relative impact of which the absolute levels prevailing in Britain overstate. At least equally important in affecting the overall demand for new cars was credit, which became increasingly restrictive during the late 1960s. The minimum deposit on cars was raised in 1965 from 20 to 25 per cent and the repayment period reduced from 36 to 30 months; in 1966 this dropped to 24 months plus 40 per cent deposit. Thereafter, fluctuations in minimum deposit levels between 40 and 25 per cent were accompanied by variations in repayment periods between two and three years [Dunnett, 1980].

Such alterations did, inevitably, affect sales, but international comparisons of demand fluctuations suggest that, as with respect to taxation, the British experience was not exceptional. Comparisons between instability of the demand for vehicles in the British market with those of Germany and the United States between 1955 and 1975 reveal a remarkable similarity [Jones and Prais, 1978]. Evidence from various sources suggests, therefore, that whatever the destabilizing effect of the British governments' fiscal, credit, and general interest rate policy measures on the motor industry, the essential characteristic of the motor vehicle as a capital good, subject to postponable purchase or replaceable from the large second-hand market, was probably the dominant influence on demand in Britain as in other countries.

There is no doubt that the effects of these policies on the industry were adverse, but it does not necessarily follow that government policies explain all, or even most, of the relative deterioration in Britain's international ranking from the 1960s. There is little doubt that the macroeconomic policies of British governments intensified instability, perhaps to a greater degree compared with the effects of policies in Europe. In all countries the market for motor vehicles showed a sensitivity to cyclical changes in the ratio of cost of acquisition to income, although there may have been differences in the degree to which government measures intensified instability [Bardou *et al.*, 1982].

While disagreement exists over the importance of macroeconomic measures as causes of the industry's increasing lack of international competitiveness, there is no dispute over the adverse effects, as yet unquantified, on companies' activities of regional economic policy introduced by the British government in the 1960s. The objective underlying geographical dispersion was to reduce unemployment in depressed areas, in Scotland, South Wales and Merseyside, where firms in growth industries were 'encouraged' to locate expansion in new plants built in those particular regions. Through the offer of subsidies and the withholding of the industrial development certificates required for planning the construction of new plants, motor manufacturing investment in Britain was funnelled away from the Midlands and the South-East, the traditional centres of motor production. It was diverted to areas remote from the main assembly plants, away from the major component suppliers and from concentrations of skilled labour (Wilks, 1984). One effect was to deter rationalization of an already geographically fragmented industry and probably, though as yet to an unmeasurable degree, to increase production costs

[Dunnett, 1980]. Dunnett argued that another adverse effect was to introduce 'an unsuitable fractious labour cohort' into the industry [Dunnett, 1980, *181*]. Workers influenced by experience in collieries and shipyards or by the prevailing culture of trade union militancy associated with those areas exacerbated labour relations in the motor industry [Dunnett, 1980].

The post-war years presented the industry with a secular expansion in demand favourable to investment and the development of overseas markets on an unprecedented scale. Whereas government policies reinforced the growth of exports, macroeconomic policies intended to achieve economy-wide objectives included specific measures which intensified demand instability, with adverse effects on industrial investment. How far these policies adequately explain the relative decline which began in the 1960s, however, is open to question. So too is the extent to which they contributed to poor labour relations, regarded by many contemporaries and subsequent historians of the industry as the key to a cumulative competitive failure.

(iii) Manufacturing systems, management and labour

Before the Second World War British manufacturers had been more successful than Ford in the home market in terms of production and market share. Marketing strategies have been adduced as a major part of the explanation for the British success in establishing its leading position in Europe. Lewchuck's analysis of corporate performance seemed unhelpful in the interwar context. It is possible, however, that Lewchuck's explanation for the roots of decline gains plausibility when applied to the period after 1939. The shift in the balance of power between labour and employers may have increased the importance of the difference emphasized by Lewchuck between the Fordist approach to mass production, involving a high degree of managerial control over labour, and the British system in which piece rate payment systems contained an inherently greater potential for a high level of control by labour over the work process [Lewchuck, 1986].

Before 1940 the potential for workers in British factories to assume quasi-managerial functions, central in Lewchuck's analysis, remained unfulfilled, the explanation for which was the weakness

60

of trade unions in the face of anti-union employers, and the growth of semi-skilled, non-unionized labour [Tolliday, 1987a]. After the war a sellers' market shifted the balance in favour of organized labour at the same time that employers' introduction of automatic transfer machines to meet rapidly expanding demand required renegotiation of piece rates. This new technology so increased the minimum efficient scale of production that in order to secure scale economies it became essential to utilize plant intensively. Increased capital–labour ratios required sharply increased labour productivity through continuous production and improved working practices [Maxcy and Silberston, 1959; Rhys, 1972].

In the final assembly process, which amounted to between 15 and 20 per cent of the average car manufacturer's costs in Britain, the economies from flow production were exhausted at an output of 100,000 units per year, and the same was true of foundry operations. However, the machining of major components – cylinder blocks for example – could be carried out most efficiently at levels between 400,000 and 500,000, utilizing very expensive, model-specific equipment. The pressing of body panels, roof and doors necessitated an output of 1m units a year for optimum production in order to obtain lower unit costs by intensive utilization of hugely expensive dies. This process, therefore, set the overall technical optimum scale for car manufacture at about 1m units in the 1950s. By 1970 further technical advances in the various processes had doubled the overall optimum figure to 2m units, though only the giant American producers – General Motors, Ford and Chrysler – were operating at these levels [Rhys, 1972].

Even at the much lower levels of output of European and British manufacturers, efficient labour utilization and a greater intensity in the use of capital became imperative from the 1950s when automatic transfer machines were widely adopted in most British factories of appreciable size [Turner, Clack and Roberts, 1967]. Under such conditions, Fordism, according to Lewchuck, had a major advantage. It assured management greater control over the deployment of workers and the pace of production. This was a critical determinant of unit costs. The abnegation of management under the 'British system' placed British companies at a serious disadvantage at a time when accelerating technical change required close and effective managerial control over work processes, particularly the pace of assembly lines and day wages. Piece rates placed that control largely in the hands of workers on the shop floor [Lewchuck, 1986].

British employers found themselves at a relative disadvantage in this respect. To varying degrees, they had favoured piece rate systems during the interwar years at a time when organized labour was weak. This was followed by a political context in which the wartime National Government and the subsequent trade union-dominated post-war Labour government were sympathetic to labour. War and the post-war boom created labour shortages and enabled workers, particularly at shop floor level, to assume an appreciable degree of managerial control.

The likelihood either of altering the existing payments system or of adapting working practices was not increased by managerial attitudes to labour. Addressing the National Union of Manufacturers in 1947 Leonard Lord, chairman and managing director of Austin, and soon to lead BMC, referred to the government's plea for cooperation between industrialists and workers and asked 'With whom are we going to cooperate – the shop stewards? The shop stewards are communists' [Wyatt, 1981, *239*]. The superior investment, productivity and financial performance of Ford after 1945 suggests that although there may be sound reasons to reject Lewchuck's hypothesis when applied to the British industry before 1939, such an analysis might offer a valid explanation of the decline of the motor industry under the very different circumstances prevailing after the war. Not only had the technology of car production altered dramatically, but the fragile state of trade unionism before the late 1930s was transformed by a growth in membership. After the war government ensured that trade unions gained some form of recognition from employers although, with the exception of Vauxhall, managers were grudging in making this concession [Clayden, 1987; Thornhill, 1986].

Recognition provided a climate conducive to a rapid growth in union membership at a time of booming demand for car workers. The well-established practice whereby trade union officials and the Engineering Employers' Federation negotiated formal wage agreements for the whole of the engineering sector, rather than for the car workers as a separate group, continued after the war. This was a system which took little account of the disequilibrium in the labour market in the fastest growing industry within the engineering sector. One effect was to make recruitment to car factories more difficult, another was to frustrate workers' material aspirations at a time when market conditions were favourable. Furthermore, although considerable changes occurred in car technology and working conditions after the war the EEF continued to refuse to negotiate on matters other than wages and hours [Turner, 1971; Durcan, McCarthy and Redman, 1983]. In response to pressure from the shop floor, the Amalgamated Engineering Union and the Transport and General Workers' Union placed a growing reliance on shop stewards to secure effective organization [Turner, 1971]. In the light of subsequent criticisms of the role of stewards in undermining industrial relations and preventing improvements in productivity it is relevant to ask how they were so successful in achieving not only substantial increases in pay but an extraordinary degree of control over the work process. Was this the main cause of the industry's deteriorating performance from the 1960s?

The consequences for factory management of these developments may be illustrated by the events at the Standard Motor Company in Coventry, which during the 1950s was the third largest British car producer. In 1949 Standard was the first company to adopt a piecework system intended to facilitate the introduction of new technology and production methods. The arrangements adopted drastically reduced the numerous grades to which piece rates applied. The complexity of the previous piece rate structure was replaced by a system whereby production bonuses were paid to large 'gangs' of workers operating as teams. In the determination of production levels, therefore, the collective responsibility of the 'gang' largely superseded that of the individual. Part of this new arrangement was the principle of 'mutuality', embodied in the 1922 Agreement but since then largely ignored by employers. Mutuality required managers to seek shop stewards' agreement on working practices before implementing change of any kind affecting production. These included not only piece rates, but manning levels, work station mobility and the setting of performance criteria. This most extreme form of managerial abdication prevailed until 1956, during a period when the growth of trade union membership in a context of booming, if fluctuating, demand for labour enabled shop stewards to enforce mutuality and to negotiate rates at levels which seriously increased costs and threatened the firm's viability.

The structure which was introduced in 1956 also included a production bonus, but by dispersing this more widely among workers Standard's managers sought to weaken the position of shop stewards, though in the event with little success [Melman, 1958]. The post-1956 agreement left the mutuality principle intact, and with it shop stewards' power to affect working practices. Stewards in other car factories pressed for similar concessions as a way of enhancing earnings and protecting conditions at work. The result was that job control became a central feature of labour relations in British car factories for more than twenty years, generating a tension [Melman, 1958] between full-time union officials constrained by agreements and procedures, and shop stewards whose resort to unofficial action embarrassed union officers [Turner et al., 1967].

Piecework accompanied by less than complete managerial control also prevailed at BMC. Evidence submitted to the Royal Commission on Trade Unions in 1966 revealed that at BMC's largest plant at Longbridge shop stewards were expected not only to avoid disputes and strikes but were also called upon to coordinate production at shop floor level. Several factors have been adduced to explain why in the 1950s and 1960s a sectionally-based labour movement succeeded in filling the vacuum left at shop floor level by the weakness of managerial control. The sheer growth in the number of trade unionists working in the industry provided the basis for strength in bargaining. The marked cyclicity of demand, in part the consequence of government monetary and fiscal policies, coupled with employers' characteristic short-term hire and fire practices led to a search by the trade unions for greater security of employment and earnings [Turner et al., 1967]. However, the form which these campaigns took was influenced not only by union strength through large membership but by multiple unionism.

The existence of numerous unions in the industry resulted in interunion competition to establish workplace organization, another factor which complicated industrial relations. So too did overlapping jurisdiction between unions which produced rivalry in recruiting members, notably between the Amalgamated Engineering Union and the Transport and General Workers Union. Both of these unions placed a growing reliance on shop stewards, regarded as essential in securing effective organization. Clayden considered this aspect of union activity in the motor factories to be an important yet neglected factor explaining the particularly vigorous development of workshop organization in the motor industry [Clayden, 1987]. He suggested that the interaction of sectional trade unionism with the employers' strategy of indirect managerial control through piece rates resulted in 'sectionalism within the work place, job control and fragmented bargaining' [Clayden, 1987, *321*]. Workers tended to be loyal to organization at gang level, much less so at plant level, and hardly influenced at all by union organization outside the factory [Adeney, 1989].

Similar strictures were made on the adverse influence of shop floor power on productivity and costs in a report by BL's joint productivity committee in the 1970s, in which the working practices of the company were compared with those of its competitors. The report drew attention to the greater amount of non-productive time, frequently prolonged by go-slows and arguments about job mobility or demarcation, more relaxation time, more generous time standards, and a worse record of disputes [Marsden *et al.*, 1985]. Just how important these factors were in contributing to the industry's difficulties, and identifying where responsibility for such practices lay, requires further consideration.

(iv) The role of organized labour: strikes and productivity

Compared both with other manufacturing industries, and with the motor industries of other countries, the British motor industry was highly strike-prone. The number of stoppages increased tenfold between 1948 and 1973 while average numbers of workers involved grew ninefold. Working days lost were seven times as many at the end of the period [Durcan *et al.*, 1983]. The rise was appreciable from 1964, reaching a peak in the 1970s. In 1969–73 the average number of strikes was 273, resulting in the loss of 1.8m working days. In 1974–8 strikes averaged 194 and working days lost numbered 1894. Figure 4 shows the long-term trends in production, strikes and associated working days lost.

There is some evidence which suggests that the percentage of man-days lost in vehicle manufacture in the 1960s and 1970s was half as high again as in the United States and ten times greater

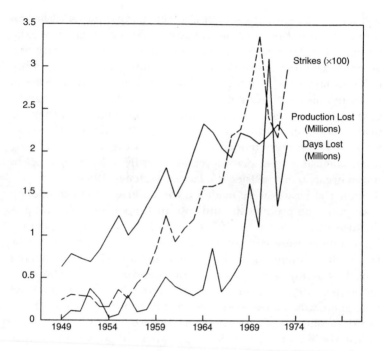

Figure 4 *Strikes, working days lost and production, 1949–74*
Sources: Turner, Clack and Roberts (1967), p. 110; Durcan, McCarthy and Redman (1983), p. 315; SMMT

than in Germany [Jones and Prais, 1978]. As the scale of plants in both countries was considerably larger than those in Britain, and as in manufacturing industry generally larger plants tended to be more strike-prone, it has been suggested that the explanation for the difference in strike propensity is to be found in contrasts between the industrial relations systems in the respective countries [Jones and Prais, 1978].

Those differences also affected the impact of strikes on production and productivity. In contrast to the United States and Germany, where single union plants entered into contracts for a specific period, British multi-union plants enjoyed much shorter periods of strike-free intervals. Jones and Prais concluded that because productivity in volume car production depended on large plant size the failure of employers and workers to resolve disputes without stoppages was the most important single factor contributing to the industry's decline [Jones and Prais, 1978]. In various forms this stress on industrial relations as the major factor appears both in contemporary commentaries and in historians' analyses [HC 14th Report, 1975; Bhaskar, 1979; Dunnett, 1980].

There is ample evidence of the weaknesses of labour relations. In 1966 the motor industry Joint Labour Council appointed Jack Scamp, a leading trade unionist outside the motor industry with a reputation for skill in negotiation and conciliation, to investigate the industry's record. Three aspects received particularly critical comment. One was the irregularity of employment and the casual attitudes it engendered among both employers and employees [Durcan, et al., 1983]; second, piece rates, though potentially productive, had an adverse effect because of the high degree of interdependence between the producers and their suppliers; third, stoppages at suppliers produced irregular earnings and differentials, both contentious issues which have led to disputes involving car workers. Yet when piecework payments halved in 1970–3 strike activity remained at high levels [Durcan et al., 1983].

While noting certain specific management weaknesses, notably the continued refusal of the EEF to bargain on matters other than wages and hours, Scamp's report reserved its major criticism for the lack of internal structure of the many unions represented in the factories. To this, more than any other single factor, was attributed the failure to enforce collective agreements which lay at the heart of the industry's appalling labour relations record. It was this fundamental weakness which has been described as evidence of 'obsolescence in institutions' [Turner et al., 1967, 339]. Dunnett and others have linked poor labour relations with associated stoppages and restrictive work practices to inefficient utilization of plant and low productivity [Dunnett, 1980; Bhaskar, 1979; Jones and Prais, 1978].

Others have played down the connection between strikes and productivity. Turner, Clack and Roberts, and Durcan, McCarthy and Redman explained the sharp rise in the level of strikes and working days lost between the late 1950s and early 1980s in terms of fluctuations in the demand for cars. Turner et al. suggested that the connection might be explained by management allowing stoppages to drag on or to widen out at times when demand fell. This strategy avoided redundancies and did not require explicit acknowledgement by negotiators on either side [Turner *et al.*, 1967]. Because this hypothesis refers to motivations it remains little more than a plausible explanation. Calculation of productivity by Jones and Prais, however, suggests that the direct effect of working days lost due to labour disputes was marginal. For example the number of vehicles produced per worker per year fluctuated between 7 and 9 not only throughout the 1950s but also during the subsequent period of very high strike activity after 1964. This gave some credibility to the Turner hypothesis [Jones and Prais, 1978].

Comparisons with other European industries suggest that the British motor industry was not exceptional in experiencing labour difficulties. Introduction of the new technology into French and Italian factories, involving a shift towards higher capital–labour ratios, new job classifications and renegotiated pay, resulted in rising absenteeism, high labour turnover and major strikes [Bardou *et al.*, 1982]. One important difference in affecting success in achieving closer control over labour on the Fordist model on the Continent is to be found in the contrasting character of the labour supply. During the 1950s British car factories offering high earnings attracted workers from other parts of Britain and the 1960s saw location of production outside the major centres of the trade in Scotland, Merseyside and South Wales. At the same time in Europe untrained, foreign or 'marginal workers' recruited to the motor industry required a completely new factory discipline to inculcate entirely novel working practices [Bardou *et al.*, 1982]. Britain's adverse record of strike activity compared with the rest of Europe needs to be set against markedly better figures for turnover and absenteeism among British workers [Durcan *et al.*, 1983].

Industrial relations were damaging to the industry. Without more systematic quantitative comparisons, however, it remains difficult to assess the relative importance in their effect on the industry of British car workers' propensity for 'formal' conflict with employers compared with the 'informal' type which characterized industrial relations in the rest of Europe.

(v) Industrial relations: Fordism and post-Fordism

How far did Ford's post-war industrial relations system contribute to that company's superior productivity record? Willman argued that in this respect Ford possessed a 'comparative innovative advantage' [Willman, 1986], a view consistent with Lewchuck's conclusion on the superiority of Fordism as a system of management control in which high day wages were coupled with firm labour discipline [Lewchuck, 1986, 1987]. Tolliday has questioned this analysis, which he regards as valid, but only to a degree, before Ford's takeover of Briggs Bodies in 1953. From that time onwards Ford began to experience similar influences to those affecting British-owned factories. Unlike workers at Dagenham, Briggs workers had been accustomed to shop floor bargaining and differential wage-grade structure. When they were moved from Southampton to man the new Dagenham Body Department a determination to perpetuate custom and practice led to a series of disputes and strikes in the face of Ford management's refusal to concede their demands [Tolliday, 1991].

The 'bell ringer' dispute of 1957, when a shop steward stopped the line to call a meeting over workers who had been disciplined by managers for leaving Dagenham to lobby a meeting of national negotiators, signalled a period of endemic conflict. The intensity of the conflict was aggravated by a group of well-organized communists, 'a union within a union', resisting managers' attempts to reimpose their right to manage [Adeney, 1989]. In 1962, a year after the parent company had purchased the remaining shares held in Britain to restore complete American ownership, the managers of Ford UK effectively destroyed shop floor organization at Dagenham after a major strike precipitated by the dismissal of 17 activists [Tolliday, 1991]. At Ford's Halewood plant near Liverpool, however, stewards established some control over manning levels and work pace. By the mid 1960s Ford's rigid wages structure began to change as competition for skilled labour compelled managers to extend special merit payments and allowances throughout the company. This resulted in differentials in addition to those built into Ford's rigid grading system.

These departures from the company's traditional 'industrial creed', in which workers and shop stewards were completely subordinate in return for high day wages, culminated in radical reform in the 1970s. When that occurred, however, it was in reaction to the company's failure to enforce in the courts a mutually binding employment contract with Ford's national Joint Negotiating Committee as part of the 1969 pay deal. Ford managers had expected to receive the support of the Labour government because of its declared aim nationally to find an alternative, orderly system of industrial relations in place of strife. Ford's attempt to devise their own legally enforceable contract proved unpopular with the trade unions, while the government gave no indication of approval for Ford's action. Opposition from the unions and inactivity of government combined to prompt a rethinking of management's approach to labour. A realization that legally enforceable contracts on the American model would not provide a method of orderly pay bargaining was followed by a growing recognition that neither would the traditional Ford managerial ideology of paternalism and direct control offer a solution to the problems of labour management [Tolliday, 1991].

These events finally prompted the move towards a new relationship with the shop floor representatives, accepting their legitimacy as negotiators, while retaining their own managerial prerogative intact. Shop stewards and plant conveners became part of the newly formed joint negotiating committee designed to integrate union organization into the collective bargaining process. Tolliday concluded that by the late 1960s – as in the mid 1950s – managerial control over the labour process was failing, the evidence of which included frequent labour disputes, off-standard performance, inflexible skills, loss of productive time and lack of right-first-time production. Time lost due to strikes rose [Tolliday, 1991].

Tolliday's account of industrial relations at Ford, therefore, reveals variation and, beginning in the mid 1960s, the gradual disintegration of the Fordist model system of labour management. Market pressures, together with the success of shop stewards in British factories in raising earnings, diminished the longstanding differentials between workers at Ford and those employed elsewhere, with adverse effects on Ford's labour relations. Coincidentally, following the formation of BLMC in 1968, moves were made to abandon piecework in British factories and to introduce machine pacing and measured daywork along traditional Fordist lines [Tolliday, 1991].

By such steps it was envisaged that British managers would restore 'the right to manage'. Measured daywork was introduced on a plant-by-plant basis between 1971 and 1974 in return for fixed earnings levels. However, shop stewards succeeded in negotiating bargaining rights with respect to effort norms, work methods and manning levels – the principle of 'mutuality'. In effect, while the responsibility for pacing and work effort was removed from the workers, with whom, to a considerable extent, it had lain under the piece rate system, managers now found themselves responsible. The extent to which managers succeeded under the new system depended not only on their ability to counteract shop floor concern over improving job conditions and security, but critically on managers' preparedness to shoulder their new role of maintaining continuity and quality of production, a question to be pursued below.

(vi) Fordist structure and strategy: the managerial organization

While the validity and importance of the contrasts between the classic Fordist system of management and labour relations and the so-called 'British system' have occasioned debate, little disagreement exists concerning the significant differences between the corporate structure and investment and marketing strategies of Ford and its British competitors in the volume car market. Ford's renaissance in the 1930s coincided with the replacement of American by British executive management including Sir Percival Perry, investment in new models, for the first time designed specifically for the British market, and the construction of the massive Dagenham plant equipped for mass production to supply cars throughout Europe.

For the post-war period, as for earlier periods, Lewchuck has stressed what he regarded as an important difference in the investment policies of American and British companies. Ford invested more heavily than British motor companies, a contrast which Lewchuck explained in terms of a preference in British companies for high dividends at the expense of capital retained in the business – itself a consequence of the readier availability of external finance [Lewchuck, 1985a, 1986]. British companies – and Vauxhall – did raise external capital after the war, but Maxcy and Silberston stressed the relative unimportance of capital issues or bank advances as sources of funds; a similar conclusion was drawn by Rhys for the 1960s. He detected no deliberate corporate policy of high distribution, although the maintenance of dividends in the face of fluctuating profits does suggest an appreciable sensitivity of managers to stock market perceptions of equity capital. At the same time, however, and in contrast to the years between 1946 and 1962, depreciation became a much more important source of funds compared with retained earnings [Rhys, 1972].

He concluded that while equity capital and bank loans were important in financing total assets during the 1960s, gross fixed investment was largely dependent on internally generated funds [Rhys, 1972]. Neither Maxcy and Silberston nor Rhys offers evidence which unambiguously supports Lewchuck's stress on British companies' preference for low profit retention, interpreted as reflecting a short-term corporate view which relegated investment and growth in their priorities. Of greater significance is the difference between the depreciation policies of Ford compared with the British companies. Between 1929 and 1938 Ford's depreciation figure was 16 per cent above that of Austin and Morris combined, a level which rose to 24 per cent between 1947 and 1956 [Maxcy and Silberston, 1959]. Although Maxcy and Silberston's analysis shows the proportion of retained earnings plus depreciation to have been greater after 1945 than before the war, the increase in Ford's net assets grew so substantially that by 1956 their value exceeded that of BMC (comprising Austin and the Nuffield organization) [Maxcy and Silberston, 1959]. Ford's fixed asset per employee (a measure of capital intensity and a proxy for machinery per worker) and output–labour ratio in 1969 were roughly three times those of the major British producer, BLMC, for which the figure was the lowest in Europe. Of the five least capitalized firms in Europe, two others, the American subsidiaries Vauxhall and Chrysler UK (formerly Rootes), were in Britain. The five firms with the greatest capitalization were in Germany [Rhys, 1972].

Although highly capitalized by British standards, Ford UK compared poorly with Ford (Germany), Opel, the General Motors subsidiary, and Volkswagen, where value added per employee was also substantially higher than that of Ford UK [Rhys, 1972]. These relativities persisted during the 1970s [Bhaskar, 1979]. The fact that Ford UK's real assets fell between 1968 and 1973 suggests that even Fordist management was not immune from the effects of government policies and labour difficulties, regardless of whether these are considered to have been of primary or only secondary importance in explaining the industry's weakness. Unlike the

British manufacturers, however, the multinationals possessed alternative options, for the reorganization of the Ford companies in Britain and Continental Europe in 1967 in the form of Ford of Europe heralded a new stage in the trend towards the globalization of motor manufacturing (pp. 107–15).

The question remains, why did Ford perform so much better than its British competitors within an environment of sluggish economic growth in an industry particularly sensitive to the effects on demand of government macroeconomic policies? Ford, like British firms, was constrained in its choice of investment location and was equally affected by taxation and credit policies, yet Ford succeeded in generating levels of profits which enabled asset ratios to increase and productivity to remain above that of British car makers.

Ford's immediate post-war success was the outcome partly of a strategy, determined in Dearborn, that the company should compete strongly in the British market. This helps to explain the considerably heavier capital investment in Britain compared with that in other European subsidiaries [Reich, 1990]. Ford benefited too, however, from especially cordial relations with the British government, a consequence, it has been suggested, of Sir Percival Perry's role in the national armaments procurement policies, and of the building of the Fordson tractor plant in Cork, seen as a valuable contribution to increasing food production in the UK [Reich, 1990; Wilkins and Hill, 1964]. In addition to favourable steel allocation, another advantage which strengthened Ford's position after the war was the £6.6m government-funded plant built in Manchester to produce Merlin aircraft engines. The investment in multi-purpose machinery, again financed by government, was adaptable to car production when the war ended, whereas the single-purpose machinery installed in many British factories for aircraft production was not [Wilkins and Hill, 1969; PEP, 1950].

Reich has argued that the goodwill generated by Ford's role during the war, together with the company's safety from the threats of nationalization when the Labour government was formed in 1945, enabled Ford to disregard price regulation, part of the government's anti-profiteering policy, although the company did observe limitations on dividend payments [Reich, 1990; Dunnett, 1980]. As a result, Ford's post-war profits increased rapidly between 1945 and the Korean War. Another important factor peculiar to Ford in 1945 was that it possessed a factory built for the purpose of mass production. In accordance with the pre-Dagenham vision of Sir Percival Perry, Dagenham had been intended to be the major Ford plant in Europe supplying Continental markets with Ford vehicles. Such a vision helps to explain the factory's annual capacity of 200,000 units whereas the maximum number of vehicles sold in the British market in any year before 1940 had been 318,000 cars and 78,000 commercial vehicles in 1937, a level not exceeded until 1953 [SMMT].

When in 1934 the control of Ford's German subsidiary was acquired by the National Socialists, the power to determine Ford strategy in the rest of Europe was also weakened, in part the effect of rising intra-European tariffs. This left Dagenham dependent almost entirely on the home market, which limited production to levels at barely half-capacity. When after the war a decision was taken in Detroit to abandon the Perry pan-European strategy, Dagenham's large capacity became at once a spur to expansion and an advantage at the higher levels of production which were to become necessary for all successful volume car producers in Europe. The immediate post-war years also saw a reform of Ford's corporate organization in preparation for the most rapid expansion in the company's history since before the First World War.

A central policy-making committee was set up in 1948 to help shape the company's activities in Britain. The new managing director was Sir Patrick Hennessey, Irish by birth and one of the three senior managers in charge of Dagenham – regarded by contemporaries as the strongest team of any of the Ford subsidiaries [Wilkins and Hill, 1964]. After taking a course in agriculture at Cork University, Hennessy had worked through the foundry, machine shops and assembly department of Ford's tractor plant, moving to sales and service at Trafford Park and on to Dagenham, where he took charge of purchasing and became general manager in 1939 [Jeremy, 1984–6]. By adopting American purchasing methods, involving interactive production planning with suppliers, Hennessey held down the cost of intermediate inputs.

In 1951 Ford managers from Dearborn introduced financial control systems already in use by the parent company. This also marked the beginning of recruitment and training of financial analysts and later of engineers from the universities and the transfer of some engineering functions from Detroit to Dagenham. A new product-planning department was established in 1953, introducing a marketing strategy to develop new models which incorporated simple but effective engineering, displaying a concern first with price and second with design; designing to a price rather than pricing a design. The first fruit of this reorganization was the Anglia, introduced in 1959, the first British Ford to provide four forward gears and sell 1m units [Adeney, 1989]. Production concentrated on a small number of models to maximize economies of scale. In order to ensure the potential profitability of each model, costs were contained by redesigning those components found to exceed their target cost [Wilkins and Hill, 1964]. The integration of the planning and organization of marketing, product development, and pricing with cost control and production became a central feature of the company's corporate culture.

(vii) Industrial structure, organization and corporate culture: the origins and performance of the British Motor Corporation

The contrast with Ford's British competitors after the war could hardly have been greater. Each of the four large producers, Austin, the Nuffield organization, Rootes and Standard, produced cars aimed at the popular 8 to 10 horsepower market, though each also produced a range of higher powered, more expensive models. The small specialist manufacturers which supplied only medium- and high-powered, expensive models included Rover, Triumph, Rolls Royce and Jaguar. The contrast between the single-plant, centralized operations of Ford at Dagenham and the fragmented, multi-company, multi-plant, multi-model characteristics of the British volume car industry was not lost on contemporaries. The 1950 PEP report criticized the continuing production of a variety and range of models which the volume of production could not economically support [PEP, 1950].

The report acknowledged that since 1939 the number of basic models in production had been reduced from 136 to 48, of which 17 came from the factories of the Big Six. Even so, the number was still considered much too large to enable either manufacturers or suppliers (whose components and accessories bought in from outside suppliers were estimated to be two-thirds of a car's factory value) to reap scale economies [PEP, 1950; NAC, 1947].

Twenty-five years later the CPRS report also pointed to the excessive number of plants as the cause, to a large extent, of the industry's weakness [CPRS, 1975].

International comparisons made more recently by Jones and Prais revealed that the number of volume car assembly plants in Britain in the mid 1970s was 13, compared with 12 in Germany [Jones and Prais, 1978]. It appears that the significant disparity lay in the size rather than in the number of plants. Census of production data suggests that by 1961 German median plants were larger than those in Britain and by the mid 1970s the output of the median German plant was estimated at between four and five times that of the British figure [Jones and Prais, 1978].

In terms of employment the contrast was less striking, mainly because of the higher productivity of German plants. In Germany the three largest car assembly plants each employed between 30,000 and 50,000 workers, not dissimilar to the workforces of the largest three American plants. The average of the biggest three in Britain was 23,000, consisting of Ford at Dagenham, 28,000, Austin(BL) at Longbridge, 22,000, and Morris (BL) at Cowley, including the nearby Pressed Steel plant, 20,000. The greatest relative divergence in plant sizes was between the vehicle assembly plants of Britain and Germany which was reflected in extreme differences in productivity. Neither in component manufacture nor in the production of bodies for commercial vehicles were significant productivity differences recorded [Jones and Prais, 1978]. Because critics of the industry's performance have dwelt on the lack of scale economies in British plants it is important to know when the discrepancy originated in order to explain this weakness. The difference with US plants pre-dated the First World War, but on the basis of census figures Jones and Prais suggested that the origins of adverse comparisons with Germany were to be found not in the immediate post-war period, when the national Advisory Council and PEP were so critical of the industry, but from the early 1960s.

A degree of structural change had occurred in the British industry in the form of ownership (see Figure 5), and in the number of models and engines in production. Even within the Nuffield group of companies, basic models in production fell from 38 before the war to 10 by 1946, and engines from 21 to 5 [PEP, 1950]. But these developments involved neither the construction

```
Austin                                          ⎤
                                                │
Morris Motors  ⎤                                │
               │                                │
Wolseley (1926)│ Nuffield        BMC            │
               ⎬ Organization    (1952)         │
MG (1930)      │ (1936)                         │
               │                 BMH            │
Riley (1938)   ⎦                 (1966)         │
                                                │
Daimler (1960) ⎤                                │
               │ Jaguar                         │
Guy (1960)     │ Group                          │
               ⎬ (1966)                         │
Coventry Climax (1963)                          │
               │                                │
Jaguar         ⎦                                │
                                                │  British
Leyland                          ⎤              │  Leyland
                                 │              │  (1968)
Rover (1967)                     │              │
                                 │              │
Standard       ⎤ Standard        │ Leyland      │
               │ Triumph         │ Motor        │
Triumph (1944) ⎦ (1961)          ⎬ Corporation  │
                                 │ (1967)       │
West Yorkshire Foundries (1945)  │              │
                                 │              │
Albion (1951)                    ⎦              │
                                                │
Crossley (1945) ⎤                               │
                │                               │
Maudsley (1945) ⎬ AVC                           │
                │                               │
AEC (1945)      ⎦                               ⎦
```

Figure 5   *Principal mergers involving British motor manufacturers before 1968 showing years of acquisition and amalgamation*

of new plants nor rationalization and concentration of production in fewer factories. Post-war conditions sustained existing companies, their physical manufacturing capacity increased by the Shadow Factories built for war production. Any explanation why, during the twenty years after the war, German plants were so much larger must also explain why the concentration of production in fewer plants did not take place in Britain, and why the range of models in production was not reduced much further. This requires an examination of the origins, organization and management of the British Motor Corporation during the 1950s and 1960s.

In 1924 market forces had produced financial difficulties for Austin and pushed Wolseley, the Vickers subsidiary, towards bankruptcy. This had prompted the irrepressible Dudley Docker, acting for Vickers, always alert to money-making opportunities from once and for all financial rationalization of manufacturing companies, to try to persuade Morris and Austin of the advantages of a merger of the three companies. Although Austin was prepared in effect to concede ownership and control to W. R. Morris the latter was opposed to even the slightest compromise to his independence. He also expressed a fear that such an organization would be 'so great that it would be difficult to control and might strangle itself' – a prescient observation, as events were to show later [Church, 1979; Davenport-Hines, 1984]. More than a quarter of a century later Lord Nuffield (formerly W. R. Morris) was still the major shareholder in the Nuffield corporate empire; but at the age of 75 he was preoccupied less with business than with philanthropic activities – although the decision to merge with Austin to form BMC he took, as was his practice, without reference to his Board of Directors [Overy, 1976].

Meanwhile, the combined market share of Ford and Vauxhall rose from 23 to 29 per cent between 1946 and 1952, the American proportion exceeding that of the two largest British volume producers. This was the context in which Nuffield finally agreed to sell out to Austin, thereby consolidating the 40 per cent share of the domestic car market in a single British company and safeguarding the organization he had originally established [Maxcy and Silberston, 1959; Overy, 1976]. BMC's relative superiority in terms of production compared with the other manufacturers in Britain is shown in Table 7. In the short term, under conditions of high levels of demand BMC's ability to exploit the company's assets fully was impressive. While the company's asset growth lagged behind that of Ford, which in 1956 exceeded the BMC's asset value, the latter produced some 50 per cent more units than Ford. This was achieved by duplicating lines and machines in existing factories rather than by constructing a new purpose-built plant

Table 7 *Shares of Car Production by Major Motor Manufacturers in Britain, 1947–89 (%)*

	1947	1954	1960	1967	1974	1978	1982	1985	1989
Austin	20.9								
Morris	19.2	38.0	36.5	34.7					
Jaguar	1.6	1.5	1.7	1.4					
Standard-Triumph	13.2	11.0	8.0	7.9					
Rover	2.7	1.7	1.6	2.7	48.2	50.0	43.2	44.4	36.0
Rootes-Chrysler	10.9	11.0	10.6	11.7	10.9	16.1	6.3	6.4	8.3
Singer	2.1								
Vauxhall	11.2	9.0	10.7	12.7	8.0	6.9	12.7	14.	16.0
Ford	15.4	27.0	30.0	28.4	25.0	26.5	34.5	30.3	29.5
Nissan									5.9

Grouping brackets (as shown in the original):
- Austin + Morris → BMC
- BMC + Jaguar → BMH
- (Standard-Triumph, Rover) → British Leyland / BL / Rover Group
- Rootes-Chrysler + Singer → Chrysler → Peugeot–Talbot

Sources: Dunnett (1980), p. 20; SMMT.

for mass production. None of BMC's ageing factories was closed during the company's lifetime.

The formation of BMC was untypical of mergers in the industry in that hitherto few had taken place between firms which did not face serious financial difficulties [Maxcy and Silberston, 1959]. Described by a former manager as 'the merger that never was' [Pagnamenta and Overy, 1984] the new organization came under the management of Leonard Lord, managing director and deputy chairman. Through a central panel Lord controlled investment allocation to the constituent companies, and to that extent reorganization represented a move towards a multi-divisional structure. His rationalization programme was limited to rearranging the production of engines (reduced in number from nine to three), axles and gear boxes to secure greater specialization between plants. However, assembly continued to be dispersed in the factories of the merged companies. This was partly because BMC continued to produce the models associated with the formerly competing companies in an attempt to retain goodwill. Dispersed production was also the longer term solution adopted to meet the rapid expansion and scale of demand for the Mini, the company's best selling model in the 1960s, which was assembled in three different locations [Adeney, 1989].

Rationalization did not extend to marketing strategy. Until the early 1960s the Nuffield and Austin organizations retained separate entities each possessing a board of directors and a set of accounts [Turner, 1971]. Such a corporate structure was not conducive to a rationalization of planning and the formulation of overall strategy. It reflected the history of managerial disarray within the Nuffield organization before 1952, which after merger was compounded by personal animosities between the managers of the merged companies. Both Leonard Lord, managing director and deputy chairman of BMC and formerly the senior figure at Longbridge, and George Harriman (who like Lord had been a senior figure first with Nuffield then with Austin), initially also deputy chairman and joint managing director of BMC, viewed the merger as a long-awaited victory for Austin [Turner, 1971].

Yet the marketing strategy continued to embrace the Nuffield philosophy of a full model range. Although the number of engines was reduced, BMC offered 14 different models, 9 from the former Nuffield factories and 5 from Longbridge. Each basic model was adapted, by variations in trim and accessories, to appeal to customer loyalties for whom the badge denoting the company of origin was regarded as an important selling advantage [Turner, 1971]. 'Badge engineering', as it became known, was symptomatic of a policy of sales competition between the constituent organizations. The senior managers continued to encourage competition between Longbridge and Cowley, the former producing vehicles under the Austin name, the latter under the Morris banner, each marketing separately through Morris and Austin dealers [Turner, 1971]. Hanks, formerly Morris's chief executive who later became vice-chairman of BMC, displayed a similar determination to that of Lord, his predecessor, to perpetuate this rivalry after merger [Adeney, 1989]. BMC was a firm divided against itself.

Distributors and dealers were retained as separate sales outlets, reinforcing the intracompany competition between the Austin and Nuffield marques which survived the merger. The first 'corporate' BMC model, the Austin 40, was introduced in 1958, yet by comparison with Ford practice product planning continued to be as unsystematic as in the past. BMC's chairman and managing director explained the company's cautious approach to product development as a consequence of the high costs of retooling necessitated by the new technology introduced in the 1950s. In his view only the superior financial resources of Ford and Vauxhall made a regular model replacement policy economic. For that reason the company concentrated on advanced engineering so as to avoid the need for frequent model replacement. This contrasted with the incremental approach to innovation which was characteristic of Ford [Turner, 1971]. Not until 1949 had Morris appointed an 'experimental engineer', and a 'standards engineer', when still the 'research director' complained that 'development thinking took place only in pubs, at dinners, and in washrooms'. Nuffield's spending on research was about 1 per cent of turnover [Overy, 1976].

Even after merger the pricing of vehicles continued to be largely an intuitive process. Lord had adopted the principle that popular Austin models should be less expensive than Morris cars competing in the same range, yet nowhere was information available on production costs specific to particular models [Adeney, 1989]. Not until 1965 was a director of planning appointed with the responsibility of pricing models with reference to market criteria. The same year brought the appointment of a finance director and the establishment of a market research department. It also marked the introduction of a policy to recruit graduates, reversing Nuffield's antipathy to training other than through apprenticeship, again in contrast with the practice at Ford [Turner, 1971; Wood, 1988].

The limited effect which these belated changes had on rationalization and restructuring before BMC was taken over was revealed in an independent report on the company undertaken in 1968 on behalf of its prospective buyer, Leyland Motors. The report stressed the lack of an effective system of financial control and product planning. It also pointed to the prevailing corporate priorities which placed engineering excellence above financial control. The report also criticized low investment [Turner, 1971].

The history of the Mini exemplified these weaknesses. The price of this commercially successful model, both in the home and overseas markets, was fixed intuitively on the assumption that only if it was the lowest priced car on the market would sales of the Mini be large. In pricing the model before the prototype had been produced, no allowance was made for the full cost of innovations incorporated in the new design by (Sir) Alec Issigonis. This design for the prototype construction of the Mini, which incorporated front-wheel drive, transverse engine mounting and high-performance engineering, was carried out under the personal control of Issigonis. Allowed virtual freedom from cost limitations, Issigonis's design was highly innovative, but the cost implications were left to solution at the production stage while pricing was fixed regardless [Turner, 1971; Wood, 1988; Adeney, 1989].

The huge sales of the Mini which immediately followed the launch in 1961 led Ford's planning department to strip one down for detailed cost analysis. On each model priced at £496 the estimated loss was £30. Over a period of years, Sir Terence Beckett, Ford's manager of product planning, correlated BMC's profits with sales of Minis and the slightly larger 1100/1300 model. An inverse relationship prompted the conclusion that BMC had assets estimated at £170m locked up in the manufacture of these two models, neither of which earned the company a return [Adeney, 1989]. Sir John Pears of Cooper Brothers, management accountants, concluded that the lack of cost information within BMC explained why the company had failed to devise a sensible way of pricing cars. This was symptomatic of a lack of managerial cohesion and of an autocratic managerial ethos which gave little sign of a sense of direction for the company other than the pursuit of engineering excellence [Turner, 1971; Wood, 1988]. Sixteen years after the formation of BMC (enlarged in 1966 by acquiring the body builders, Pressed Steel Fisher, and by merging with Jaguar to form British Motor Holdings) the accounting system in Britain's largest vehicle manufacturer was still not designed to supply precise information on model costs [Turner, 1971].

Between 1945 and 1968 car production in Britain fell from first to second place behind Germany. The growth in industrial concentration affected ownership rather than managerial structure, producing an underdevelopment of the strategic managerial function, one tangible manifestation of which was badge engineering, a symptom of intracompany competition, minimal central cost control and a lack of planning. Underdevelopment of the management function was also in evidence at the shop floor level, where various forms of piece rate systems enabled labour to shift the balance of advantage against managers. These continued to hamper the company during the 1960s, when the move towards freer international trade began to intensify competition in home and overseas markets. It was this development which precipitated further mergers, culminating in 1968 with the birth of British Leyland.

3 The Vicissitudes and Collapse of a 'National Champion'

(i) Anatomy of a merger: the rise of British Leyland

When in 1968 BMH merged with the Leyland Motor Company to form the British Leyland Motor Corporation (BLMC, renamed BL in 1977), the new organization became the second largest motor manufacturer outside the United States and the fifth largest manufacturing company in the private sector in Britain. At the beginning of the second phase of decline, Britain's 'national champion' held 40 per cent of home sales and 35 per cent of the truck market [Cowling, 1980]. The effects of mergers have been a subject for debate between those who stress the greater opportunities mergers offer for achieving economies of scale and scope – the prevailing conventional wisdom within government in the mid 1960s – and those who emphasize the contingent nature of these potential advantages. This raises the questions: Why did the merger occur? What were its effects on the industry's performance, and how can they be explained?

The possibility of restructuring and reorganizing the entire industry became stronger from the 1950s when Leyland Motors, the small Lancashire manufacturer of specialist commercial vehicles, acquired others of a similar kind: Albion Motors and Scammell, followed by ACV in 1962 [Turner, 1971]. Leyland's success in this sector contrasted with its history of failure as a car manufacturer. For a brief period during the 1920s Leyland had produced two models, the 'Straight Eight', designed to compete with Rolls Royce, and the Trojan, a small, solid-wheeled vehicle whose tram-wide wheel span could create a hazard for urban driving. In 1961 Leyland re-entered the car trade when it took over Standard Triumph, followed in 1967 by Rover. Both were manufacturers of specialist cars, neither of which was strong financially but each dominated market niches, Standard Triumph in the market for sports cars, Rover selling medium-quality cars in addition to the unique Land Rover for commercial use. In these limited market niches the Leyland–Standard Triumph connection proved relatively successful during the six years before the Leyland–BMH merger in 1968 [Turner, 1971; Rhys, 1972].

The ultimate decision on merger was that of the shareholders of the participating companies, among whom Prudential Insurance held a substantial interest, following lengthy negotiations which involved the respective company directors of Leyland and BMH. However, the Labour government played a key role in bringing about the regrouping. The initial suggestion that the two companies might merge came from Anthony Wedgwood Benn, the Minister of Technology in the Labour government of Harold Wilson. In preparing a national economic plan civil servants at the Department of Economic Affairs had identified the car industry as one of the economy's major problems, which led both Benn and Wilson to become personally involved in its solution through merger. Formal pressure to promote the prospective merger was applied through the Industrial Reorganization Corporation (IRC). This was the governmental body (of which Sir Donald Stokes, Leyland's managing director, was a member) charged with the responsibility for promoting mergers for the purpose of rationalizing British industry [Turner, 1971].

The IRC ensured that in the restructuring which resulted, 'cress', the codename for BMH, which was regarded as a badly-managed company, should be subordinate to 'mustard', the corresponding cipher for Leyland Motors, whose contrastingly strong financial position derived mainly from the buoyant commercial vehicle sector. The IRC assumed that Sir Donald Stokes, regarded as one of the country's leading salesmen, would be able to duplicate that company's success as a producer of cars for the volume car market. The government's intervention at that time betrayed increasing concern at the growing strength of the American multinational manufacturers in the British market. The Leyland merger, described by Benn as 'a fantastic achievement' [Benn, 1988, 16] may thus be seen at least in part to have been driven by government in response to Chrysler's acquisition in 1967 of a substantial shareholding in the ailing Rootes group, the other British volume car producer [Young and Hood, 1977].

In order to understand why the merger failed it is necessary to establish the objectives of the various parties to the negotiations. The government's aim was to bring to an end the fragmented structure of the industry in order to consolidate and rationalize an independent British industry. As the largest manufacturing exporter, the industry was important to the balance of payments, which before devaluation in 1967 had been especially critical to the government's economic policies. As a major employer, too, the industry attracted special interest from the increasingly interventionist Labour administration [Turner, 1971]. Sir Donald Stokes and his co-directors expressed concern that their shareholders' interests should not suffer as a result of the merger, a fear which diminished somewhat as the relative share prices of the two companies moved in opposite directions, lowering the cost to Leyland when the takeover occurred.

The managers of BMH were initially opposed to merger. Even after personal intervention by the Prime Minister to persuade its ailing chairman to agree in the national interest, the BMH managers proceeded on a mistaken assumption that despite the company's deteriorating financial position the larger size of the volume car business would ensure the dominance of BMH in any amalgamation and the subsequent corporate reorganization [Turner, 1971; Adeney, 1989]. As it turned out, the financial strength of Leyland together with the connivance of the IRC enabled Leyland to take over BMH and to control the board of the new company. Furthermore, an immediate £25m loan from the IRC for retooling was the first indication that Leyland was to be afforded special treatment as a 'national champion'. The following year Stokes became deputy chairman of the IRC, the organization responsible for monitoring the performance of British Leyland – of which he was now the head [Turner, 1971].

Outright takeover of this kind was virtually without precedent in the industry since the 1920s. Hitherto, the holding company had been the characteristic form of amalgamation. By perpetuating the constituent corporate entities, and with few exceptions retaining the owners or managers who ran them, holding companies had not been instrumental in achieving corporate and industrial integration. The formation of the British Leyland Corporation was seen as offering an opportunity to begin the long postponed transformation.

Expressed in terms of the historian's model which relates successful business strategy to structural adaptation [Chandler, 1990; Channon, 1973], a necessary, though not sufficient, precondition for effective corporate performance was a transition from a personal and entrepreneurial structure, characteristic of both BMH and Leyland, to a corporate, managerial organization – in reality as well as on paper. Ostensibly, Lord Stokes's task was to integrate eight public companies into one, but the constituent companies were themselves the products of previous mergers [Cowling, 1986]. The chronological history of the precursors of British Leyland is presented in Figure 5. Lord Stokes later reflected on the overwhelming difficulty he had faced in welding together '50 or 60 different companies all trying to retain their independence even though they had been taken over' [Adeney, 1989, *254*]. John Barber, Leyland's new finance director, formerly of Ford and the electrical manufacturer AEI, referred to BLMC as consisting of 'a mass of unrationalized plants and unrationalized products' [Adeney, 1989, *256*]. On its formation BLMC produced roughly one-half of all cars made in Britain (see Table 7).

(ii) The effects of merger

In 1974–5 three official reports offered diagnoses of the industry. The first was prepared by an independent body, the Ryder Committee, set up by the Labour government as a preliminary to deciding on a policy for the industry. The second was that of the Trade and Industry Sub-Committee which conducted its investigations, complete with extensive evidence from numerous witnesses from within the industry, in response to the Ryder Report and its recommendations. The third was produced by the government's 'think tank', the Central Policy Review Staff (CPRS).

The conclusions drawn by these three bodies differed mainly in emphasis, though they were in agreement that BLMC was in danger of terminal collapse. Ryder emphasized the handicap of the company's outdated plant and machinery, its excessively centralized corporate organization, and also criticized weaknesses in management–labour communications which contributed to a bad record of industrial relations. The Trade and Industry Sub-Committee was also critical of a debilitating bargaining structure, but attached greater importance to overmanning and excess capacity which reflected a failure to respond to consumer demand [HC, 14th Report]. The CPRS levelled similar criticisms, but blamed lack of maintenance of plant and equipment and working practices which slowed workpace and restricted productivity. With various differing emphases, these several interpretations of the nature of the company's problems have been incorporated in subsequent attempts to explain decline [Bhaskar, 1979, 1983; Dunnett, 1980; Williams *et al.*, 1983, 1987; Wood, 1988; Adeney, 1989].

While the reports differed in specifics they revealed general agreement on the extent to which the government's optimism concerning the future of BLMC in 1968 had not been realized. Why not? One reason was the sheer scale of the rationalization needed. Britain's only large vehicle manufacturer now possessed 48 factories which included 23 major plants accumulated by BMH. The product range ran from low-priced volume cars, specialist, luxury and sports cars to heavy goods vehicles, buses, fire engines and dustcarts. A former assistant controller of Ford of Europe who was recruited to British Leyland described the characteristics of the new merger as a 'multiplicity of style, multiplicity of technology, multiplicity of everything ... The organization lacked a common system, a common ethos or culture' [Wood, 1988, *247*].

The attempt to provide a coordinating multi-divisional structure to a considerable extent perpetuated historical continuities and functions, which existing geographical dispersion tended to reinforce. The Truck and Bus Division was based at Leyland; Austin–Morris, the volume car division, produced in Birmingham and Oxford; while the Specialist Car Division included the Coventry factories of Triumph and Jaguar, and the Solihull Rover plant. The remaining four divisions were those specializing in bodies, power and transmission, parts and kits for assembly and an overseas division dealing with exports and foreign subsidiaries [Salmon, 1975]. The activities of these divisions, each represented by a managing director responsible for the profit centre and participating in the formulation of overall corporate policy along with senior staff directors under the chairman, were subject to control by central office organized on a basis of staff and line management. A large central staff department was responsible for monitoring and controlling the seven divisions. A greatly enlarged staff, however, was not accompanied by the delegation which was normally a feature of the managerial hierarchies of multi-divisional companies and this led to duplication. No fewer than 21 directors and managers reported directly to the chairman, in effect perpetuating the personal, hierarchical tradition of the company he had worked for since boyhood [Salmon, 1975]. A former manager employed by Ford, later briefly chief executive of BLMC, described the process of decision-making at BLMC as lacking in documentation and almost intuitive. The company lacked explicit corporate objectives, which meant that neither monitoring nor analysis of corporate performance were systematically employed in the assessment and formulation of business strategy [Pagnamenta and Overy, 1984].

While a multi-divisional structure had been introduced, the strategic policy-making, coordinating, monitoring and informational advantages which restructuring was intended to achieve remained unrealized. Six years after the merger the Specialist Car Division remained largely unchanged, evidence of a continuing conflict of subcorporate philosophies compounded by a legacy of distrust rooted in the past [Salmon, 1975]. This continued to be the case after the Ryder reforms of 1975. Although they were intended to reduce centralization by substituting four for seven divisions (combining Austin Morris and Specialist Cars to form BLMC's car product planning division), they left the balance of power shared between divisional managing directors and the overall corporate managing director as a matter of 'personal chemistry', and left lower level local management within the divisions with ill-defined authority [Salmon, 1975; HC, 14th Report, 1975].

Clashes of culture continued to occur. One example is that which led to the closure of a high-technology £25m plant built at Solihull to produce a new Rover model, codenamed SDI, to compete with BMW. The insistence by Leyland head office on rapid quantity production of the new car conflicted with Rover's tradition of emphasizing high-quality vehicles manufactured in relatively small volume. Difficulties over overmanning, demarcation, and overoptimistic sales forecasts culminated in BLMC's senior managers overruling the Rover factory's inspectors. The first immediate consequence was a poor reputation for the new car; the second was the plant's closure after only five years [Pagnamenta and Overy, 1984].

The evident failure of the Leyland version of multi-divisional operations did little to integrate the former corporate units and provoked suspicion and hostility towards central management. This has been interpreted as evidence of Leyland managers' lack of experience of mass production, large workforces, and extensive managerial hierarchies [Cowling, 1986]. Whereas at the time of the merger Leyland was making 23,000 commercial vehicles a year, the car-making capacity of BMH exceeded a million, and employed 200,000 workers [Wood, 1988]. The chairman and managing director of the new company, Sir Donald Stokes, had worked for the small, Lancashire-based heavy truck and bus-making company since he joined as an engineering apprentice in the 1930s, although it was as an effective sales manager for Leyland heavy trucks and buses that he was to make his contribution to the company's success after the war. From its formation as the Lancashire Steam Motor Company in the 1890s until 1963, three generations of the owner-founding Spurrier family had been chairman, or general manager, or both [Turner, 1971]. When in 1963 terminal illness removed the ageing Henry Spurrier II from Leyland, and Stokes became deputy chairman and managing director of the Leyland Motor Corporation (the precursor of BLMC), he was the only director below the age of 70, reflecting a disregard for managerial succession characteristic of British firms [Turner, 1971].

Rootes, the second largest British car manufacturer and the greatest loss-making motor company in Britain, had been in the hands of the Rootes family since its foundation in the 1920s. Until 1967, when control was conceded to Chrysler, the firm had displayed characteristics similar to those of Leyland. It had been built up through a series of mergers. An incoming Chrysler manager described the Rootes organization as having been divided into a number of largely uncoordinated separate family empires. The autocratic rule of Lord Rootes, who died in 1964, had been followed by 'a period of less decisiveness', but most significant as evidence of the intuitive approach to managing the organization was the lack of adequate cost accounting and other financial controls and the absence of a profit plan [Young and Hood, 1977; Wood, 1988].

In an attempt to keep Rootes in British hands, in 1967 the Minister for Technology, Anthony Wedgwood Benn, had invited both BMH and Leyland to rescue Rootes, then sliding towards bankruptcy [Benn, 1988]. Neither company expressed interest, perhaps because the directors of both companies recognized a lack of managerial depth in their respective organizations. BMC's graduate recruitment policy was of recent origin and had suffered high turnover. Stokes was also conscious of Leyland's own serious shortage of managers capable of transforming the British car industry [Adeney, 1989]. Later he described himself and other Leyland managers as not being 'motor car people.... We switched off lights and saved pennies' [Wood, 1988, *163*].

Clearly the supply and quality of management was a weakness which the departure of certain senior managers of high quality, resulting from the takeover in 1968, probably made worse [Turner, 1971]. In part this was because the production and market characteristics in the heavy commercial vehicle sector contrasted markedly with the mass produced car sector. New cars produced since the Second World War were of unitary body shell construction providing the rigidity to support the power train. This, together with the importance of body shape and size as the major source of product diversity, meant that body panel construction dominated the economics of car production. In the heavy commercial vehicle sector chassis and body remained separate, the special requirements of industrial consumers resulting in a high degree of product diversity [Rhys, 1972].

In the light and medium weight section of the industry, van and truck production benefited from standardization and since the 1930s had been dominated by the mass producers, Morris, Ford and Bedford (Vauxhall's CV subsidiary). Until the merger with Leyland, BMC was the largest CV manufacturer, responsible for about one-third of total CV production, although that was concentrated almost entirely on light, car-derivative models and medium vehicles. Ford and Bedford each accounted for between 20 and 25 per cent during the same period. Leyland's concentration at the opposite end of the range gave that company barely a 6 per cent share of CV production, which on merger with BMC rose to about 40 per cent. Until that time Leyland's success depended mainly on small-scale, skilled, labour-intensive manufacture of vehicles to customers' detailed specifications in a segment of the market less affected by the extreme slumps experienced by the car industry [Rhys, 1972].

The failure of the Leyland merger of 1968 was the result of both external and internal factors. International movements towards freer trade intensified the competition from manufacturers in Continental Europe. Internal weaknesses were those of an industry in which historical and personal continuities frustrated corporate reform along multi-divisional lines. A complicating factor was that the volume car division came under the management of those whose training and experience was mainly in the production and sale of specialist commercial vehicles and, briefly, medium-quality cars. Furthermore, BLMC was also constrained by the indirect effects of various governments' balance of payments, income and employment policies, which inhibited labour relations and created a climate of uncertainty unfavourable to sustained investment policy. Employment policy was the government's justification for rescuing Rootes, thereby intensifying competition within the home market and reducing Leyland's opportunity to gain volume and reduce costs. Whatever the corporate weaknesses, government policies hindered rather than helped the British motor industry.

(iii) British Leyland's productivity dilemma: markets and productivity

After 1968 car production accounted for 70 per cent of BLMC's business, though the models in production were becoming outdated. The Mini, still the top-selling model, was nine years old, and the Morris 1100/1300 six years old. Both were under competition from more recent models, particularly the Ford Cortina (replaced by the Escort in 1968) and the Vauxhall Viva [Cowling, 1986; Williams, Williams and Haslam, 1987]. Yet BLMC lacked the financial resources to design and launch a completely new model. The Austin Allegro, which went into production in 1973 to replace the Morris 1100/1300, closely resembled its predecessor, while the Morris Marina, first introduced in 1971 to compete in the medium-car class, incorporated some of the parts used for the 1948 Morris Minor [Adeney, 1989].

The inability to sell models in large numbers was critical to production at competitive costs. This was Leyland's productivity dilemma. Estimates of the minimum production levels at which the scope for economies of scale existed suggest that only in the assembly process were some models produced in sufficient numbers annually (between 0.2m and 0.25m) to achieve them. The optimum levels of annual output for engine blocks of about 1m and for body panels of 2m (at which level all processes achieved scale economies) were beyond those of any model produced in Europe, with the exception of the Volkswagen 'Beetle' [Rhys, 1972]). Some progress was made towards the rationalization of models, reduced from 39 to 24; however, the extent of this process was limited by the board's reluctance to concentrate production in fewer factories with a smaller workforce, regardless of the criticisms of BMH reported to the Leyland directors prior to new merger [Cowling, 1986].

BLMC's new finance director, formerly a manager at Ford and subsequently at AEI, considered the company to be overmanned by 30,000, which Stokes attempted to reduce through exhaustive consultation with the trade unions. The recent massive redundancies which had followed the rationalization of the GEC–AEI merger had alerted the trade unions to the possibility of a similar threat, increasing the risk of further strikes should BLMC pursue a similar tactic [Wood, 1988]. A disinclination on the part of management to tackle the overmanning problem at a time of difficult labour relations meant that it was not until 1978, when the factory at Speke was closed by Stokes's successor, Michael Edwardes, that plant rationalization and a reduction of manning levels began in earnest.

Not surprisingly, Cowling's estimates of efficiency gains for this period show none attributable to merger [Cowling, 1986]. Furthermore, such model rationalization that did take place had damaging consequences for the company's share of the home market. Ford had reduced its dealer network in 1963 by one-third over five years, at a time before imports had gained a permanent and substantial foothold. When BLMC carried out a similar rationalization between 1968 and 1976 one-third of the 7000 dealers whose franchises were withdrawn soon took up franchises for importing European and Japanese vehicles. Even after rationalization Ford dealers sold on average double the number of cars of each outlet handling BL cars [Williams, Williams and Thomas, 1983; Adeney, 1989].

Leyland's delay in rationalization was a serious handicap during the 1970s when the state of the motor industry worldwide, damaged by the oil crises, their economic ramifications and the emergence of excess capacity in the industry, made profitable trading problematical. Until the late 1960s none of the major companies in Britain had recorded more than one or two losses since the war, though Ford's profit record was considerably superior to that of the other volume car makers. Between 1960 and 1967 the pre-tax profits of BMC-BMH amounted to £132m compared with Ford's

£174m [Dunnett, 1980]. Ford, moreover, produced roughly one-third fewer cars than the British manufacturer. Higher unit profits and a higher rate of return on capital were the key factors explaining Ford's financial strength and its superior cash flow which supported higher gross investment [Maxcy and Silberston, 1959; Silberston, 1965; Rhys, 1972; Dunnett, 1980]. Between 1974 and 1982 Ford was the only car manufacturer in Britain which did not make a loss (after tax) in every year except one. Before tax, the comparable figures for BL show a net loss over the period of £198,000, although losses were widespread also among companies in Europe and in the US [Bhaskar, 1984a].

The origin and timing of the British company's weakness may also be highlighted by comparing estimates of the profitability of the three principal categories of motor vehicles: volume cars, specialist models (which before the Leyland merger were produced by Jaguar, Rover and Standard Triumph) and trucks and buses. Estimates suggest that during the early 1960s BMC's volume car production generated roughly three times as much profit as each of the other two categories. Between 1967 and 1972 the Austin–Morris Division of BLMC contributed 17 per cent, the Special Cars Division 49 per cent, and trucks and buses 34 per cent [Bhaskar, 1975]. Thus, well before the sharpest reductions in tariffs occurred and the globalization of international production which followed, volume car production was already the least profitable of BLMC's activities, and exercised a financial drag on investment in the other two divisions.

The difficulties which the combined effects of these developments presented to BLMC's market position intensified as a result of two further changes. One was the shift from indirect to direct full-line competition, the commencement of which Williams et al. [1987] identified in 1977. Ford entered the small car market for the first time with the Fiesta, followed by BL's attack on all three main market segments: small, light and medium. The other new factor was Vauxhall's successful re-emergence as a major car maker in Britain in the 1980s, when the company produced the J Series Cavalier. Financed by GM and designed by its Opel subsidiary, the model was aimed at the medium fleet car market [Williams et al., 1987]. Market fragmentation and direct competition led to a relatively even distribution of sales of the ten best-selling models, four from Ford, three each from Vauxhall and Rover, none of which reached 10 per cent of the share in its class. Under such circumstances, capturing 20 per cent of the market with three models became virtually impossible for any company. In effect the possibility of reaching scale economies without very substantial exports or a multinational operational basis had been removed [Williams *et al.*, 1987].

BLMC's relatively poor sales record has been attributed in part to the distinctive structure of the market for cars in Britain which Williams *et al.* [1987] have argued made the industry particularly vulnerable to aggressive importing when tariffs fell. Their analysis revealed two almost discrete submarkets, one for business users, either as purchasers of fleets or separately, and the other for individuals making private purchases. The submarket for business began to develop in the mid 1960s when employers reacted to government income policies by offering subsidized company cars as a way of circumventing pay constraints for managers. Between 1964 and 1975 cars owned by firms and subsidized by employers (who also received tax concessions on their purchases) [Ashworth, Kay and Sharpe, 1982] rose from 17 per cent of domestic sales to 40 per cent. This meant that the segment of the market which was expanding most rapidly was that which BLMC was least well equipped to exploit, for Ford already dominated sales in this market with the medium-sized, medium-priced 1500–1600 h.p. Cortina, its capacious boot and rakish lines contrasting with those of the Maxi and the Marina [Wood, 1988]. Another consequence of this structural change was that BLMC was left with an extensive distribution network suited to private consumers. Yet when it was rationalized the beneficiaries also included importers, who from the early 1970s were attacking the private car market and required established outlets through which to distribute sales [Williams *et al.*, 1987].

The context in which restructuring and rationalization were attempted became increasingly difficult from the mid 1960s. The sheer acceleration of changes affecting competition in both home and overseas markets tested entrepreneurial and managerial limits to the full. After nearly 50 years of protection the tariff was reduced to 25 per cent in 1963, to 17 per cent in 1969 and to 11 per cent by 1972 (22 per cent on CVs). From 8.3 per cent on entry to the EEC in 1973, by 1977 the tariff had disappeared altogether. These developments coincided with the worldwide recession which resulted from the threefold increase in oil prices in 1973–4, and again in 1979, intensifying competitive pressures. The nature of that competition also became more complex as the process of globalization, initiated by the American multinationals in the late 1960s, began to alter the dynamics of international trade in motor vehicles. As tariffs fell, the British industry became vulnerable to competition from Europe, particularly from the high-volume, high-productivity Volkswagen plant. In 1965 the German industry had produced 6.4 'equivalent' motor vehicles per employee per year compared with 5.8 in Britain; by 1970 the comparable figures were 7.5 and 5.6, and in 1976, 7.9 and 5.5 [Jones and Prais, 1978]. Successful new designs from the French state-owned Renault company also contributed to import penetration [Rhys, 1972].

A further development requiring flexible managerial organization, additional capital expenditure and highly trained staff was the higher rate of innovation made possible by computer-assisted design, engineering and production. The demand for such technological and human resources was further increased by new safety and pollution standards required for vehicles sold in the American market, a trend which during the 1980s became both international and progressive and caused difficulties [Whipp and Clark, 1986], although by that time the British industry's strong trading position had already been severely eroded.

Freer international trade, faster productivity growth in Europe and Japan, together with the greater competitiveness of foreign cars in terms of price and design, enabled importers to penetrate the British market. When during the peak sales and production period during the summer of 1971 easier credit and lower purchase tax on cars had the effect of increasing car sales by 43 per cent, imports were sucked in to meet a demand which the domestic industry could not supply. This episode has been described as the catalyst which heralded a permanent shift in the tastes of British consumers in favour of foreign cars [Dunnett, 1980]. Thereafter, regardless of the buoyancy or otherwise of the British economy, import penetration took a rapidly rising share of the home market (see Figure 6).

The effect of import penetration was compounded by the relative decline in Britain's share of world exports, which since the 1920s until the 1950s had been the largest of all European countries. Thereafter, although export volumes rose, they represented

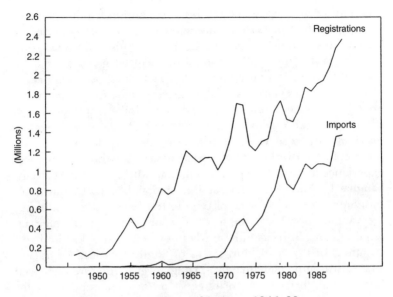

Figure 6 *New car registration and imports, 1946–89*
Source: SMMT

a lower proportion of output. Trade with the major Commonwealth markets which had been a sizeable importer of cars from
Britain before 1939 continued to offer a potential offset against
restricted exporting to protected Continental markets, both
before and after the creation of the EEC [Jones, 1981]. Australia
and New Zealand accounted for nearly 40 per cent of cars
exported from Britain between 1947 and 1956, when quota restrictions aimed at developing an indigenous manufacturing industry
resulted in a sharp drop in exports to Australia. Part of this loss
was for a time compensated by exports to the United States, where
Volkswagen had already established a niche market, selling the
small European utility vehicle as a second family car. In 1959 British cars comprised nearly 50 per cent of American car imports, but
they fell sharply to 10 per cent in 1961 when American manufacturers countered with their own new range of compact cars.
Among European manufacturers only Volkswagen retained a substantial hold on this market [Rhys, 1972; Nelson, 1967].
    Meanwhile, the adverse implication for trade of British manu-
facturers' relative absence, historically, from the European market

was reinforced by very marked trade expansion within the EEC. Whereas by the late 1970s France and Germany traded on average 1.25m units within Europe, the figure for British sales was below 200,000. Part of the explanation for the British lag lies in the existence of Continental tariffs, but other factors are also important. One was the pressure exerted by government for a very rapid growth in exports during the post-war dollar shortage which meant that both quality control and the building of an adequate distribution and service network took a lower priority than production for an overheated market [Rhys, 1972; Wood, 1988]. The experience of the widely publicized, new 'world car', the Standard Vanguard, aimed especially at international markets, illustrates this. Since there was a lack of time to test for foreign conditions, the Vanguard proved to be too underpowered to meet American requirements, insufficiently sturdy to overcome the dust and corrugations of the roads and tracks of South America, Australasia and Africa, and inadequately sprung to give comfort on paved Continental surfaces. Accordingly it attracted a reputation for unreliability which became difficult to dispel [Turner, 1971; Pagnamenta and Overy, 1984].

Neither in France nor Germany did a dealer network selling British cars exist during the 1950s. Furthermore, when in the 1960s successful sales of the Mini offered an opportunity to develop an international distribution system (following Volkswagen's strategy) BMC had opted for a policy of local assembly, exporting kits rather than complete vehicles. Of the three assembly operations set up by BMC in Belgium, Italy and Spain in the 1970s, however, only the Belgian plant at Seneffe was owned by BMC. The licences issued to the Italian and Spanish companies for the assembly and sale of Minis and Allegros were merely temporary, however, terminable either by lessor or licensee. Only the highly innovatory Mini, with transverse engine and front-wheel drive, obtained success in Europe, where in the 1960s it accounted for 70 per cent of all of British car sales [Williams et al., 1987]. From the 1970s, however, these innovations had been incorporated as common features of international car design.

European manufacturers had introduced newly designed models which included the slightly larger, less Spartan varieties of super-Minis and hatchbacks to compete with the Mini and with BLMC's much less successful models, the Marina and the Allegro.

Replacement of these, however, was prevented by BLMC's low profitability and insufficient cash flow in consequence of limited sales volume and underpricing. This was worsened by continued dividend payments which at least in the short term weakened the overall financial position of the merger. It is also possible that one effect of dividend payments during the recession of 1966–7, the intention of which was to shore up BMC's market value [Turner, 1971; Rhys, 1972], was to inflate Leyland shareholders' expectations after the merger. This might explain why between 1968 and 1975 out of £74m profit £70m was paid out to shareholders, evidence of that short termism which Lewchuck considered to have been a long-established feature of British motor manufacturers. However, state and industry discussions of the level of investment needed to rescue BLMC in 1975 make it clear that the scale of the company's diversion of profit was small in comparison with the volume of funds required to introduce new models and methods. It should not, therefore, be regarded as fundamental to the company's deep-seated problems [Adeney, 1989].

One of the consequences of BLMC's weakness in the market at a time when vehicle manufacturers in all countries were experiencing the effects of a contracting world demand was a dramatic rise in import penetration as vehicle producers looked to overseas markets to compensate for sharp falls in home demand. In 1975 imported cars accounted for 33 per cent of new registrations, exceeding BLMC's share of 31 per cent for the first time. BLMC's production dropped from 1.7m in 1973 to 1.26m in 1975. In the same year the company declared a loss of £76m, representing between £90 and £160 on every vehicle sold. When the banks refused to continue their financial support it was left to the government to decide whether to allow the 'national champion' to collapse, accepting the dire consequences for unemployment and the balance of trade.

(iv) The nationalized champion: policies and personalities

BLMC was not allowed to go bankrupt because the government was concerned over the effects on the balance of trade and on employment. Prime Minister Harold Wilson described the industry as 'an essential part of the United Kingdom economic base'

[Wilson, 1979, *137*]. The company was given a £50m guarantee, pending an investigation of its affairs by a small committee under the chairmanship of Sir Don Ryder, the head of the newly established National Enterprise Board (NEB). The outcome was a recovery plan agreed between government and BLMC which involved the acquisition of public equity and an expenditure of £2000m over eight years. Benn, minister at the Board of Trade and Industry, set out in detail the purpose of the government's strategy: 'to subsidize British motor manufacturing that would otherwise be replaced by imports from the EEC and elsewhere; and to provide long-term capital at non-commercial rates to re-equip and improve capacity and competitiveness' [Benn, 1989, *364*]. After Ryder's plan had been accepted by government the body responsible for its implementation was the NEB. In effect, because of the very large volume of funds which the plan called for, the British-owned industry came under state control.

The plan adopted for the industry originated from an internal company document prepared by BLMC managers even before Ryder had conducted his own investigation, but of which he became informed in the course of his inquiry. He accepted their predictions for future sales, a critical planning assumption which in the event proved excessively optimistic. It was on that basis, however, that the Ryder plan justified the need for expanding production and no redundancies. In the light of concern expressed by a cabinet minister that if the company was not rescued 'a comprehensive motor industry' and nearly a million jobs would be lost [Castle, 1980, *374*], the two assumptions underlying Ryder's strategy were at least consistent with the government's overriding aims in nationalizing the company.

They may also have been decisive in securing cabinet approval for such massive public investment. Cabinet accepted Ryder's analysis, which persuaded members that the problems, in the words of one participant in the discussion, were those of 'poor management ... grotty machinery and bad industrial relations' [Castle, 1980, *374-5*]. Part of the formula for revitalization was a change in management. Stokes assumed an honorary role and received a peerage; John Barber was sacked. The new managing director, Alec Park, hitherto Leyland's finance director, assumed the major responsibility together with a part-time chairman [Adeney, 1989].

From the outset relations between management and the NEB caused difficulties, for while the NEB held responsibility for monitoring the company's progress, reporting back to the Secretary of State, the plan was framed in such vague terms that the point of ultimate responsibility was unclear [Jones, 1981]. An intervening period ensued during which Ryder, acting almost independently of the advice from members of the NEB, not only monitored the company's activities but intervened in day-to-day decision-making. This *de facto* assumption of the role of quasi-chairman of the board was an arrangement which neither managers nor trade union leaders found satisfactory [Bhaskar, 1979; Adeney, 1989].

Equally unsatisfactory was the new system of industrial relations machinery introduced on Ryder's recommendation. On the principle of industrial democracy a joint management and trade union consultation process was established at three levels. While the principle of industrial democracy was intended to signal to the trade unions a new era, in practice the unions found the dispersion of consultation did nothing to strengthen the powers of negotiation of official trade union leaders with corporate management, consequently in 1979 they withdrew from the new Joint Management Councils [Willman and Winch, 1987]. Within two years of the new regime the consequences of an inflation rate above 20 per cent, repeated stoppages, declining sales, and an import penetration which rose to 45 per cent in 1977 brought about a further managerial reconstruction.

The prevailing ambiguity of authority was replaced by the unambiguous managerial control of a single figure who was both chairman and managing director, receiving advice from two close aides [Bhaskar, 1983]. Under a new Labour Prime Minister, James Callaghan, and a new Secretary of Trade and Industry, Eric Varley, Stokes's successor was a self-styled apostle of free enterprise. He was Sir Michael Edwardes, a member of the NEB, who was appointed on a five-year secondment from his position as chief executive of Chloride, the largest battery-supplier to the motor industry. Trained as a lawyer, though experienced in rationalizing the battery-making industry in southern Africa and therefore familiar with the motor trade, his particular strength was considered to be in managing organizational change. Closely monitored by government, but without intervention in the implementation of mutually agreed objectives, the Edwardes strategy of producing new models depended on continuing, substantial government finance. The 1975 formula worked out by the government guided by Ryder had made that conditional on three objectives – the ending of overmanning, a transformation of working practices, and improvements in industrial relations.

Under Ryder changes in industrial relations had been structural rather than substantive at the operational level [Willman and Winch, 1987] and made no significant contribution to achieving an improved strike record and reductions in overmanning. From 1977, under a new regime which no longer ruled out plant closures, Edwardes introduced a confrontational approach to the trade unions, particularly to shop floor activists, over pay claims, plant closures and work practices [Willman and Winch, 1987]. Edwardes sacked the communist Derek Robinson, popularly known as 'Red Robbo', who led the opposition against the introduction of new working practices of any kind, particularly those which allowed the movement of workers between jobs. By this action Edwardes sought to end the buying out by managers of restrictive custom and practice [Edwardes, 1983]. Cuts in the workforce of some 50,000 between 1977 and 1982, when those on the payroll numbered 83,000, were in part achieved through brinkmanship. Explaining the company's difficulties, and the need to persuade the government that the recovery strategy was working, Edwardes appealed directly and personally to meetings of shop stewards or other workers, who were asked to accept the need to reduce the numbers employed and to avoid stoppages [Willman and Winch, 1987; Adeney, 1989].

Corporate organization and structure were transformed. The car-making division, known as BL, was separated from Leyland Trucks and Leyland Buses. Unipart was the division manufacturing spare parts. With the aid of psychological testing the company undertook a systematic weeding of managerial staff at all levels, accompanied by highly selective promotion and recruitment of new senior managers, several of whom were attracted from Ford. The entire process represented a systematic attempt aimed at creating a completely new corporate culture in the boardroom, in offices and on the shop floor. These reforms proved sufficient to secure issue of £450m in government equity to enable the company to create two new models. However, the lead time for design and for the installation of new computer-aided technology left a potential two-to-three year gap between the launch of a replacement for the Mini and a new car aimed at the medium-price range.

To bridge the gap BL entered into an agreement with Honda, whose car production was similar in scale to that of the British company, and which was the third largest Japanese motor manufacturer. The first fruit of the memorandum of understanding between the companies was the appearance in 1981 of the Honda Ballade, developed in Japan, built at Canley and sold under the Triumph badge. The second product of joint Anglo-Japanese development was the Rover 800, the first of a new Rover series introduced in 1984 [Wood, 1988]. The first new all-British model from BL was the Metro, conceived as the 'super Mini', which for the first two years after its launch in 1981, manufactured in a new robotized body shop at Longbridge, sold at an annual rate of over 170,000. That was followed in 1983 by the Maestro, the new medium-sized saloon car, which also exceeded 100,000 sales. The outcome of systematic research and design and of cooperation with Honda, these developments were seen by academic commentators and many other observers as symptomatic of, as well as instrumental in achieving, a turning-point in the recovery of BL [Jones, 1985; Willman and Winch, 1987].

(v) From nationalization to privatization

A detailed re-evaluation in 1987 suggested that these judgements were misplaced [Williams *et al.*, 1987]. While the undoubted changes in industrial relations and more flexible working practices were achieved between 1977 and 1985 and acknowledged [Willman and Winch, 1987; Williams *et al.*, 1987], yet BL's falling market share at home and declining exports were signs that higher productivity was not enough to retrieve the company's position (see Table 8). Factories at home and overseas were closed in order to reduce fixed overhead costs, including the Belgian plant which had assembled over 50 per cent of the company's cars sold in EEC countries between 1976 and 1980. The closure in Belgium almost coincided with the drop in trade with Italy following Italia Innocenti's switch to Japanese engines and gearboxes for its rebodied Mini, which in the late 1970s accounted for between a quarter and

Table 8   *British Leyland Production, Exports and Share of New Car Registrations, 1970–89*

	Production of cars (000s)	Exports of cars (000s)	Total UK car sales (000s)	BL home car sales (000s)	UK market share (%)
1970	788.7	368.4	1076.9	410.4	38.1
1971	868.7	385.8	1285.6	516.3	40.2
1972	916.2	347.3	1637.8	542.4	33.1
1973	875.8	348.0	1661.4	529.6	31.8
1974	738.5	322.5	1268.8	415.4	32.7
1975	605.1	256.7	1194.1	368.7	30.8
1976	687.9	320.8	1285.6	352.7	27.4
1977	651.0	293.3	1323.5	322.1	24.3
1978	611.6	247.9	1591.9	373.8	23.5
1979	503.8	200.2	1716.3	337.0	19.6
1980	395.8	157.8	1513.8	275.8	18.2
1981	413.4	126.2	1484.7	285.1	19.2
1982	405.1	133.9	1555.0	277.3	17.8
1983	473.3	118.3	1791.7	332.7	18.6
1984	383.3	78.6	1749.7	312.1	17.8
1985	465.1	112.8	1832.4	328.0	17.9
1986	404.5	123.6	1882.5	297.5	15.8
1987	471.5	165.7	2013.7	301.8	14.9
1988	474.7	137.0	2215.6	332.6	15.0
1989	466.6	138.5	2300.9	312.3	13.6

Source: Williams, Williams and Haslam (1987), Table B2, p. 126; SMMT.

a third of exports to the EEC [Williams *et al.*, 1989]. Because the Austin cars shipped to Belgium and Italy were assembled locally they had established a status of quasi-indigenous products; without that exports to these two countries, and to Europe, collapsed after 1981. In the absence of a distribution network selling identifiably British cars, a switch to direct exports of complete vehicles required investment, successful models, and time for development. Meanwhile the high and rising value of sterling both undermined competitiveness and squeezed profit margins.

Williams, Williams and Haslam argued that BL's trading weakness was due mainly to a flawed business strategy, which resulted in a 'market-led failure' [1987, 67]. The new cars, the Metro and Maestro, were not sold in sufficient numbers, consequently between 1980 and 1985 capacity utilization averaged 56 per cent – well below the estimated 70–80 per cent needed to enable the Longbridge and Cowley factories to break even after covering depreciation, let alone secure economies of scale. Indirectly this drag on productivity was attributable, they maintained, to a failure to recognize the changing character of competition in the market, particularly within Britain. BL's strategy was the same which had led to Ford's strength in the British market since the late 1930s, the production of a narrow range of four basic models. However, because of the high capital costs of changeover each of the BL models was planned to have a long production life [Williams *et al.*, 1987]. This was in the BMC tradition and in contrast to Ford. However, even as BL's new model strategy began to be implemented, Ford introduced the small Fiesta car in 1977, while Volkswagen replaced the 'Beetle' with the more conventionally designed and extremely successful Polo. Together they heralded a new phase of competition which in certain respects resembled the full-line model price competition of the 1930s, when manufacturers divided the market into niches. This made BL's lack of new models in the medium range at once a serious handicap, for the new competitive conditions reduced the likelihood of Metro sales reaching the predicted levels on which the Metro's planned production, costing, pricing and profitability were based.

Williams *et al.* [1987] argued that BL's difficulty not only stemmed from an inability to reach production levels which would bring scale economies, but that the particular capital-intensive technology employed on the Metro line at Longbridge was disadvantageous. While it offered flexibility in producing variants of models the technology did not allow variation in the number of basic models, small and light/medium cars which could be produced on the lines. The Dagenham plant both facilitated the production of variants and allowed flexibility in product mix. The new technology did increase BL's labour productivity, but capital productivity fell. So did the profits shown in the company's balance sheets (but which actually concealed losses) [Williams *et al.*, 1987]. By 1987 the car division of BL (by that time renamed Rover

Group) held only 15 per cent of the home market; imports had risen from roughly one-third at the time of nationalization to one-half (Table 8).

In the commercial vehicle branch, too, the British company was being squeezed. Both the American multinationals and some European manufacturers were competing effectively in the market not only for light- and middle-weight trucks but for heavy commercial vehicles, the demand for which grew in response to increasing transcontinental road transport. This was a sector in which until the 1960s Leyland, and later BLMC, had performed strongly both at home and overseas [Rhys, 1972]. The penetration by European manufacturers into a market hitherto insulated from foreign competition was the result partly of the virtual stagnation in Leyland's commercial vehicle production capacity at a time when the demand, particularly for heavy vehicles, was growing, and partly to freer trade, culminating in entry to the EEC [Rhys, 1972]. Arguably the merger of Leyland with BMH into which £3bn had been injected by government between 1975 and 1988 not only failed to remedy the weaknesses of the latter but ultimately undermined the success of the former [Adeney, 1989].

Public expenditure on such a scale under a government committed to privatization was the context in which the final stage in the history of the 'national champion' began. A further collapse of BL's market share and the prospect of further losses led to the intervention of the Department of Trade and Industry. In 1986 the company's part-time chairman, and the two senior managers who had taken on Edwardes's executive role after his return to Chloride at the end of his contract in 1983, were replaced. Combining both roles, on the Edwardes model, the DTI appointed another professional manager, the Canadian Graham Day, formerly chief executive of the heavy-loss-making British Shipbuilders. It was under Day's management that British Shipbuilders had been rationalized in preparation for privatization, which had become a key feature of Conservative government policy [Adeney, 1989].

Day proceeded to carry out a similar strategy to fulfil government objectives for the British motor industry. Jaguar, the small semi-luxury car maker, had already been sold on the Stock Exchange in 1984, although initially the government had retained a 'golden share' (revoked shortly after) to protect the company from corporate, especially foreign, raiders. In 1987 Leyland Trucks was sold to DAF, the Dutch CV manufacturer, while the bus division was disposed of in a management buyout, as was Unipart. In 1985 informal, secret meetings involving executives of Ford, BL and cabinet ministers explored merger. However, a political crisis, precipitated by the sale of the loss-making Westland Helicopters, raised the issue of foreign ownership of strategic British companies, creating a climate which torpedoed secret discussions involving Ford and senior political figures close to government. A resurgence of national feeling destroyed the possibility of a merger of the British car-making division with Ford. This was a development sought by the senior management at Dagenham anxious to withstand increasing import penetration. In particular they feared the growing threat posed not only by Japanese imports but in the longer term by inward investment and car production in Britain by Japanese companies [Adeney, 1989].

In 1988 Rover Group was sold to British Aerospace, the defence and property conglomerate, which itself comprised several former state-owned mergers between problem companies. The price of £150m was widely regarded as a gross undervaluation of the company's assets (and fell well below the figures mooted in the discussions involving Ford in 1985). The £520m cash injection from government was considered to be an overgenerous dowry, while tax concessions on Rover's losses and other concealed payments strengthened suspicions that political considerations were paramount in deciding Rover's fate [Adeney, 1989; *Independent on Sunday*, 9 August 1992]. Three years later, in a conglomerate whose other divisions were also recording losses, Rover's market share in Britain slumped to no more than 13 per cent, while production was barely 0.4m and the workforce numbered 35,000.

(vi) Globalization and the role of multinationals

The ultimate demise of car and truck production by independent British-owned manufacturing companies was as remarkable for the speed with which the process occurred as for the scale and impact of that decline. That rapidity owned much to a combination of freer international trade and the growth of international competition. The effect on BLMC of rising import penetration on market share is shown in Table 8. From the early 1970s penetration by Japanese imports accounted for a small but rising share of new registrations, which led to an increasing concern among manufacturers in Britain. The result was an informal agreement between the SMMT and the Japanese manufacturers' association, whose members then held 9 per cent of the British car market, that a policy of 'prudent marketing' should limit imports voluntarily to no more than 11 per cent, an arrangement which continued to be observed (at the upper level) throughout the 1980s [Adeney, 1989].

Japanese imports exacerbated, rather than created, problems for the industry in Britain. Indirectly, however, the phenomenon of Japanese industrial expansion did have a considerable impact at a time when the British industry was at its most vulnerable. Built up rapidly during the 1960s behind tariffs at prohibitive levels, by the end of the decade Japan had become the world's second largest producer of motor vehicles. It consisted mainly of commercial vehicles and small cars exported on an increasing scale to the US, the Far East and Australia [Rhys, 1972]. The response of the three American multinationals to Japanese competition in the US market was to increase investment in overseas subsidiaries and to move towards greater international integration of their operations. One effect of this was to intensify competition in Britain and Europe.

Beginning with the formation of Ford of Europe in 1967, Ford's policy – later emulated by General Motors and Chrysler – began as an exercise intended to achieve greater cooperation between the company's operations in Britain and Germany. Ten years later the strategy had evolved into a policy of regional integration of production and marketing based on international division of labour on a global scale. In Europe, meanwhile, the 1970s saw an increased level of industrial concentration, notably with the merger of Peugeot with Citroën. In 1979 the two largest motor manufacturers were General Motors (8.8m) and Ford (5.5m), but Japanese producers were third, fifth and tenth, including Toyota (3.0m) as third largest, followed by Volkswagen–Audi. The two largest French manufacturers, Peugeot–Citroën and Renault, were among the top ten, followed by Fiat (1.4m); British Leyland (0.6m) was fourteenth [Bardou et al., 1982].

What were the effects of globalization on the motor industry in Britain? Its manifestations included a greater standardization of model range and the allocation of vehicle and component production between subsidiaries, based on comparative transaction costs. For example, the Ford Escort, launched in 1969, was manufactured both in Britain and Germany, although it was the Capri, introduced in 1977, which was the first Ford car to be conceived at the outset as a European car, built in cooperation between Ford plants in Britain and Continental Europe. In the 1980s the Fiesta, planned jointly by an Anglo-German team seconded from Dagenham and Cologne, exemplified the increasing levels of regional integration in progress. It was assembled in Germany and Spain; engine blocks were produced at Dagenham and carburettors in Northern Ireland. The proportion of locally-manufactured components in the three countries varied [Wood, 1988].

When General Motors began to rationalize production in 1979 car production and marketing were concentrated at Opel in Germany, and the manufacture of commercial vehicles in Britain. When in 1984 General Motors launched the J car (in Britain the Cavalier II), it comprised an engine made in Australia, either an automatic gearbox made in the US or a manual one made in Japan, a carburettor made in France, and pressings from Germany. The British contribution came to little more than oil filters, glass and wheels. The British content amounted to about 60 per cent, counting labour costs as roughly one-half [Adeney, 1989]. With the power train manufactured elsewhere, as far as General Motors' operations are concerned by the early 1980s Britain had become little more than an assembly centre.

The trend towards 'sourcing' seriously affects the interpretation of trade statistics. Car imports from the EEC as a proportion of new owner sales in Britain rose from 20 per cent in 1974 to 38 per cent ten years later, and exceeded 40 per cent by the late 1980s; but this was largely attributable to British-based American assemblers obtaining an increasing proportion of their car sales in Britain supplied from plants in Europe. The combined figure for these tied imports by Ford and General Motors increased during that period from 1 to 22 per cent. Thus, whereas in 1973 the MNEs were net exporters of 200,000 cars, by the mid 1980s they had become net importers of around 350,000 [Jones, 1985; Bhaskar, 1984a] (see Table 9). Their strength in the British market since the 1970s was based as much on their role as the largest importers of Ford and General Motors vehicles sold as 'British' cars as on their superiority as manufacturers.

Table 9 *Home Car Sales, Tied Imports and Exports by US Multinationals, 1976–89 (thousand units)*

	Ford			Vauxhall		
	UK Sales	Tied imports	Exports	UK Sales	Tied imports	Exports
1976	325	29	108			
1977	340	87	132			
1978	392	138	102			
1979	486	237	130	141	46	
1980	465	217	85	133	51	
1981	459	203	82	127	59	
1982	474	230	61	182	103	
1983	518	240	32	262	139	
1984	487	208	17	283	165	
1985	486	214	14	303	169	
1986	515	303	6	285	125	
1987	580	177	8	271	86	5
1988	584	249	7	304	119	1
1989	609	238	3	350	147	2

Sources: Bhaskar (1984a), pp. 529, 591, and SMMT.

Furthermore, because of the American companies' practice of sourcing, the trade statistics understate the growth in imports. Whereas in 1973 Chrysler UK models sold in Britain embodied 97 per cent local content by value, by 1983 (as Talbot) the figure was 30 per cent. Over the same period the local content of Vauxhall cars fell from 98 to 22 per cent, compared with Ford's reduction from 88 to 22 per cent. By taking local content into account Jones calculated a tied car and component import figure of 4 per cent in 1974 which rose to 31 per cent ten years later. He estimated a 'true import content' figure for the British market in 1984 to have risen to 66 per cent, compared with official car import figures which showed stability at 57 per cent between 1979 and 1984 [Jones, 1985]. The fall in exports by multinational subsidiaries was the consequence of corporate policies designed to limit intra-European trade by permitting plants in one country to export only to specific foreign markets. At the same time, the MNEs increased the export of kits for assembly in receiving countries. By 1983 Ford's net trade deficit from Britain reached £700m, that of General Motors £600m [Jones, 1985]. Therefore, at a time when the British industry was struggling to maintain a falling market share at home, the Trojan horse effect of tied imports and heavily-sourced cars which only appeared to be British was thus partly responsible for Leyland's inability to generate sales at levels which would support scale economies in production. From a national standpoint the MNEs were worsening Britain's trading balance on cars, which from 1975 went into the red.

One further indirect effect of the increasing strength of multinationals was that their reliance for basic research and development remained outside Britain. In this respect, throughout the 1970s the limited expenditure on R & D undertaken by BLMC/BL also helped to put Britain well behind France and especially Germany, which benefited from R & D expenditure by American multinationals [Jones, 1983]. Comparisons with spending in the US and Japan show expenditure on research and development in Britain to have been a fraction of the sums committed in those countries. The same is true for the number of scientists, engineers and other workers employed in motor-related R & D. The figure for Germany was 75 per cent greater than that for the UK, compared with a 10 per cent British lag behind France [Jones, 1983]. If the investment of capital and human resources is taken to reflect the strength of each country's capacity for design engineering and the development of product technology, this evidence offers support to Whipp and Clark's contention, although it is disputed by Willman and Winch, that weaknesses in these respects hindered competitiveness [Whipp and Clark, 1986; Willman and Winch, 1987].

Was the impact of multinationals entirely negative? In assessing the effects of MNEs on the British industry it is important to consider their record of investment and employment creation (or maintenance), as well as their marketing policies. For while as traders the MNEs' policies were damaging to the British industry and to the balance of trade, Ford's record of investment based on higher unit profitability compared well with that of BLMC/BL, dependent on public expenditure to offset losses to provide net investment (see Table 10). On a rising trend, between 1973 and

Table 10   *Net Profits after Tax, and Capital Expenditure of Major Vehicle Producers (£m) 1970–1982*

	British Leyland		Ford		Vauxhall		Chrysler Peugeot/Talbot	
	Profit	Cap. exp.	Profit	Cap. exp.	Profit	Cap. exp.	Profit	Cap. exp.
1970	–6	67	16	68	–9.4	11	–11	15
1971	16	50	–17	49	2.6	15	1	5
1972	22	42	28	32	–4.1	10	2	5
1973	26	63	28	42	–4	18	4	7
1974	–24	108	32	52	–18	32	–18	2
1975	–124	92	52	52	–13	14	–36	2
1976	44	114	7	56	–2	14	–43	2
1977	52	149	59	81	–2	9	–22	18
1978	–38	233	116	163	2	32	–20	7
1979	–145	259	144	334	–31	32	–41	14
1980	–536	284	347	324	–83	22	–102	10
1981	–497	201	204	280	–57	12	–91	8
1982	–293	230	165	398	–39	12	–55	4

Source: Bhaskar (1984a), Vol. I, Table 8.1; Vol. II, Tables 9.13, 10.12, 11.5.

1982 capital expenditure in Ford UK roughly equalled that of the British vehicle manufacturer. It was triple the investment made by Ford in other European countries and twice that undertaken in other subsidiaries outside the US [Bhaskar, 1984a].

State subsidies were extended to Ford and Vauxhall, mainly through regional development programmes and investment grants in the form of interest relief. These implied a tacit recognition by government that on balance the employment provided by Ford's investment outweighed the effects of tied imports and sourcing. Of the £182m spent on the plant constructed at Bridgend in South Wales in 1972, about 40 per cent is estimated to have come from public funds. Thereafter subsidies were smaller, amounting to £177m for Ford and Vauxhall between 1976 and 1983. In 1976–9, however, government injected £162m into Chrysler's heavily loss-making Talbot subsidiary in an attempt to avoid unemployment [Bhaskar, 1984b]. Unsuccessful as a policy, this measure also ran counter to the government's strategy of reducing the number of producers and promoting rationalization [Bhaskar, 1983; Young and Hood, 1977].

BLMC/BL was not the only company adversely affected by tied imports and sourcing. The components branch of the motor industry supplied parts and accessories which accounted for 65 per cent of the material cost of BLMC/BL cars, as well as some proportion which made up 70 per cent of the material costs of Fords, 85 per cent of Vauxhalls and 71 per cent of Chrysler cars. Several of the firms which dominated the market – Lucas, Chloride, Dunlop, GKN, Automotive Products, and Automotive Engineering – virtually monopolized the home market for particular components, but the trend towards restricted vehicle exports and tied imports by the American multinationals stimulated a drive for exports by the component manufacturers.

In the 1970s the deficit on the balance of trade in cars was more than offset by the surplus from components. However, the fall in BL's production coupled with a rise in imported components by the MNEs led to a marked contraction of the market. One reaction of British component producers was to undertake direct investment in production facilities on the Continent, the source of imports, in order to protect sales and retrieve profitability. Even so, by the mid 1980s the balance of trade in motor components was also in deficit [Carr, 1990]. Assessing the overall effects of the activities of multinationals, one historian's conclusion was that 'the deleterious effects of a sharp rise in tied imports, a decline in exports, a drastic diminution in local content and a reduction in car manufacturing capacity combined to accelerate the drift towards the industry's decline' [Church, 1986].

The economics of the international division of labour on which the MNEs' policies of the 1970s and 1980s were predicated had the effect of strengthening the industry in those countries where a combination of productivity, labour costs, exchange rates and profitability favoured location or relocation of production, although subsidies offered by governments seeking to attract investment were added to the equation. A comparison between Ford UK's performance and that of other Ford companies in Europe suggests that overall these factors placed the industry in Britain at a disadvantage during the 1970s and early 1980s. By European standards the performance of Ford UK was no more than average [Bhaskar, 1983]. The productivity gap between Ford factories in Britain compared with those in Europe was due, in the opinion of Ford managers, not to differences between actual line speeds or production standards, which were similar, but in large measure to interruptions on the line. Compared with European plants, British factories recorded higher rates of industrial disputes, greater relaxation allowances, and permitted more non-productive time, which was often exceeded. A particular contrast was that BL had higher manning levels and lower work standards than European plants [Marsden, Morris, Willman and Wood, 1985].

Quality differences were also identified. In the mid 1970s Ford Fiestas and Escorts made in Britain attracted roughly twice the number of complaints concerning the same models built in Germany, while repair rates on the Cortina/Taunus were half as much again for the British-made cars than for those manufactured in Belgium. In 1981 a consumer survey of warranties revealed similar disparities between the quality and reliability of cars made in Britain with those made in Germany and especially in Japan [Bhaskar, 1979, 1983]. Warranty costs of the new Metro and Maestro models exceeded those of comparable Ford and Vauxhall cars, and were ten times those of Nissan cars [Williams *et al.*, 1987]. In a period when the quality of motor vehicles, as of other consumer goods, was subjected to increasing scrutiny by magazines and journals aimed specifically at consumers, and when increasingly systematic international comparisons promoted informed consumer choice, the weaknesses in British production created further difficulties for the marketing of British cars [Bhaskar, 1983].

Weaknesses specific to the British industry were exacerbated by a slowing down in the rate of growth of world vehicle production [Bardou *et al.*, 1982]. The rapid rise of Japan to become the world's largest vehicle manufacturer in the 1980s, and an associated intensification of international competition and globalization of production added to the problem. The contraction of output from British factories resulted in merger, nationalization, and finally privatization. In 1988 Rover became a small and peripheral division of a large British conglomerate which, because of the financial arrangements entered into by the British government, the European Commission decreed must not be sold off by British Aerospace before 1993.

These developments coincided with a change in the equation affecting Britain's comparative advantage as a location for vehicle manufacture by foreign firms. The movement of the exchange rate, the more peaceable climate of industrial relations, a consequence partly of legislation and partly of recession, and the continuance of higher price and profit margins inside the British market, combined to encourage inward investment, which successive British governments actively.sought. For Japanese manufacturers these advantages added to the prospect of access to the single European market planned for 1993 and made Britain an attractive focus for investment. The eclipse of the dwindling British motor industry meant, therefore, that more than ever the ability to translate comparative into competitive advantage in Britain came to depend on the financial resources, design and production technology, managerial methods, working practices and approaches to industrial relations conducted by American, French, and especially Japanese multinationals.

(vii) Explaining decline

By themselves none of the explanations which have been offered for decline, whether they emphasize adverse government policies, obstructive and militant labour, or organizational and managerial weaknesses, adequately account for the decline of the British motor industry. Each is necessary to appreciate the complex chronology and interaction between these various factors. Lewchuck's institutional approach, which stressed the system of atomistic competition and the centrality of the effort bargain to production, has been criticized for being too deterministic by oversimplifying the relationship between institutions and human behaviour. The model offers valuable insights into the industry's immediate post-war history, underlining the erosion of managerial control which continued after the BMC merger in 1952. At the same time, however, the sheer scale of the merger offered managers an opportunity to retrieve a measure of the shop floor control conceded during the war and its immediate aftermath.

The IRC's success in achieving an effective restructuring of the electrical engineering industry, which began with the merger of GEC with AEI in 1967, contrasts with the failure of the Leyland

115

merger [Cowling, 1980]. This not only points to the IRC's lack of power to enforce managerial change beyond the structural reform which resulted in the Leyland merger, but also underlines missed opportunities within the organization. Whereas overmanning and inefficient working practices were tackled immediately at GEC they remained a problem for BLMC. Although managers recognized the need for similar, though more far-ranging, changes, shop floor control in the car factories hindered managers' moves in that direction. By reducing further the complications presented by intercompany competition with respect to wages and conditions, the greater industrial concentration in the 1960s, when the British industry became virtually a monopoly, removed one of the difficulties of reforming the effort bargain. Another major barrier to change, which perpetuated the adverse effects of piecework payments, was the prevailing disarticulated system of industrial relations. In effect, negotiations were complicated by employers being at one remove from the trade unions, and shop stewards at one remove from both. None the less, the survival of these institutional arrangements until their reform under the confrontational Edwardes management beginning in the late 1970s must be attributed partly to human failures, of managers and labour leaders.

Those who have emphasized the role of the trade unions and shop floor militancy in contributing to underutilization of plant, slow-speed working and strikes, notably Bhaskar, and Jones and Prais, likewise overlook the difficulties presented to Continental manufacturers by workers who were also uncooperative, though in different ways. Again this suggests that the effects on motor manufacturers in Britain of disruptive workers and of restrictive practices were those of degree. However, a comparison of industrial relations in German motor companies with those in Britain suggests that the differences had effects beyond those of lost production and lack of cooperation on the shop floor. The British system of industrial relations generated levels of uncertainty missing from the German system, with consequences for corporate planning beyond labour management.

Volkswagen was exceptional, in that after the Allies relinquished control the company remained in public ownership until the early 1960s. Thereafter the Federal Republic retained 20 per cent ownership, Lower Saxony held 15 per cent and the Volkswagen Foundation 5 per cent. Beginning in 1952, the Works Constitution Act and the Collective Agreements Act were two pillars of the system of 'co-determination', designed to ensure that workers were represented on the supervisory boards in the ratio of one to three management representatives, increased in the 1970s to almost one in two, and that the law should be central to industrial relations. Even during the early post-war years the public company adopted informal arrangements not dissimilar to full co-determination as the basis for relations between managers and workers. Democratically elected Works Councils were given a legal monopoly of workers' representation, who used their rights under co-determination to ensure the recruitment of trade unionists. As trade unions (and shop stewards whom the unions attempted to insert into the structure of industrial relations in the late 1950s) lacked the legal status of Works Councils, the Councils maintained their key position.

Among the consequences of this system which gave legally enforceable workers' representation at workplace, and later at company, level, was single union and company bargaining, centralized decision-making between Works Councils and companies, and an absence of interunion rivalry. From labour's standpoint there was some complaint that within this structure trade unionists were unable to exploit their position on supervisory and management boards to extract the most from employers [Streeck, 1984]. Yet the ordered structure of industrial relations in the German car industry contrasted with the unregulated, individualistic character of industrial relations in Britain. Here the complexity and volatility of industrial relations were the combined result of centralized power held by an employers' association representing other engineering industries, and competing trade unions whose power, and that of workers generally, depended heavily on the exercise of shop floor action and strikes.

A key difference compared with the German system, and an important debilitating factor affecting the British industry, was an orderliness in relations between groups and institutions embedded in law. Such arrangements enabled companies to anticipate and plan in the knowledge that the law ruled out unpredictable, precipitate industrial action. The contrasting legacy of adversarial industrial relations in Britain generated both uncertainty and mutual distrust, reducing the likelihood of cooperation to effect internally initiated changes in management strategies. These were postponed until virtual bankruptcy, government intervention, and a relaxation of political constraints (though within financial control and targets) on managers recruited from outside the organization signalled a further repudiation of the long-established corporate culture.

Especially during the 1960s government, too, contributed to the uncertainty which discouraged fundamental long-term managerial changes. However, the emphasis by Bhaskar, Dunnett, and Pollard on the role of government in the industry's decline offers only a partial explanation. Evidence from international comparisons suggests that macroeconomic policies exacerbated, rather than caused, instability. Between 1968 and 1977, however, when government assumed a central role, political considerations affected the industry's development directly. The original strategy for BLMC, initiated by government through the IRC in 1968, failed partly because the choice of company and chief executive to lead the merger lacked the experience and the managerial support needed to transform the industry. Weaknesses here stemmed in part from a lack of professional and analytical skills, but important too were the prevailing assumptions and beliefs about what was possible, as well as the limited means to achieve objectives. Neither directors nor managers at Leyland had sought or planned for a restructuring of the entire British motor vehicle industry. The failure of strategy, in which government willed merger and rationalization while leaving implementation entirely to the new company in a political climate hostile to redundancies, illustrates the weaknesses of what Wilks called 'institutional insularity' [Wilks, 1990].

Wilks argued that throughout the post-war period government policies were the outcome of government–industry relations which were vitiated by the absence of institutional linkages, reflecting a British tradition of limited public authority in industrial affairs and a respect for the autonomy of business organizations. Reich concluded that undercapitalization, a high strike rate and poor management all resulted from the form that government policy took after 1945. International comparisons, he argued, showed that the critical difference between Britain and Germany, Italy and to a lesser extent France, is that only in Britain did liberal ideology survive the Second World War. Fascism and war on the Continent fostered discriminating ideologies and institutions, allowing scope for government intervention to promote national, economic and political interests. On the other hand, in Britain intervention by the state, when it occurred, was both constrained and non-discriminatory.

By treating American companies in Britain on equal terms with British-owned companies, Reich argues, government encouraged unrestricted inflows of foreign capital which were detrimental to the competitive position of the British industry. Furthermore, he argues that British government policies based on egalitarian principles created an advantage for the American subsidiaries. The allocation of steel, during the post-war period and at the time of the Korean War, favoured Ford, and later, the government underwrote Chrysler's losses in an attempted rescue. The American firms also possessed the option of withdrawing from the industry in Britain, and during the immediate post-war period this enabled them to disregard government advice on price control and profits. Reich suggests that the compliance of British firms was influenced by a perceived threat of nationalization, while later their approach to restructuring was softened by managers' interpretations of the government's willingness to rescue the Chrysler subsidiary as a reassuring sign for the survival of the British industry [Reich, 1990].

Most of Reich's new evidence to support his model, which breaks down in the 1960s, refers to the 1940s and early 1950s. Then Ford undoubtedly benefited from the prevailing liberal ideology at that time, but it is difficult to conclude that in a period of rapid growth in demand this was to the serious disadvantage of the British industry. Ford's post-war recovery and performance were spectacular, but from the mid 1930s that company possessed the critical advantages of a large, modern production plant, which had yet to produce at full car-making capacity during peacetime. Ford also possessed depth and ability in local management, and benefited from the resources for development, skills and experience of the parent company to call upon. Moreover, one reason why Ford proved to be so competitive within Britain whereas Volkswagen outpaced Ford in Germany was the corporate strategy chosen by managers in Dearborn, who at the end of the war regarded Ford in Britain as the European flagship. Ford's share of total car production in Britain reached 30 per cent; none the less, the superiority of productivity in the industry as a whole compared with European producers until the early 1960s suggests that productivity in British factories was not substantially different from that of Ford up to that time.

A much stronger case has been made by Wilks, who argues that from the 1960s the liberal ideology, which implied 'national treatment' for foreign subsidiaries, led government to rescue Chrysler and to remain passive when other American subsidiaries resisted government pressure to act in accordance with government employment and income policies that were constraining British companies' activities [Wilks, 1990]. He concluded that the willingness to bail out Chrysler 'placed the future of the industry at least implicitly in the hands of the multinationals' [Wilks, 1990, *186*]. This is a valid inference viewed in the light of BLMC's corporate weakness and the wider institutional structure of government including the civil service. Even after the crisis of 1975 heavy public investment in the company stopped short of extending discriminatory support comparable to that received by the national champions of other major Continental car manufacturing countries. While the state came to control the British industry in 1975, and since 1968 had acknowledged the importance of maintaining a British industry, tangible manifestation of government support was limited to providing funds for investment. The institutional dealings between firms and the Departments of Trade and Industry, Environment, Transport, Regional Development and the Treasury were conducted as 'transactions between insular elites'. Wilks concluded that the lack of contact between a secretive, non-specialist civil service on the one hand, and on the other uninformed managers, ineffectually represented at the industry level by the SMMT, resulted in minimal information flows and maximum propensity for mutual misunderstanding [Wilks, 1990].

One effect of this lack of awareness of the respective concerns, priorities and objectives was to preclude what has been called 'the politics of reciprocal consent' [Samuels, 1990], crucial to which was the need generally to ensure the cooperation of those whom the government sought to regulate. Instead, government–industry relations were based on mutual suspicion and the politics of reciprocal incomprehension. Wilks distinguished between two levels at which industry–government relationships operated between 1975 and 1985, neither of which involved more than financial support and monitoring. One was the bureaucratic level, at which he described the relationship as 'distanced', the other was the 'arbitrary' relationship which characterized contacts at the ministerial level. The defects of communication and understanding at both levels contributed to the confusion between political and industrial objectives when BLMC came under state control in 1975. The government accepted a strategy which, contrary to much of the evidence available, seemed to promise industrial revival through investment, improved industrial relations, some rationalization and minimal unemployment. The government subordinated economic criteria in defining industrial success and how to achieve it, to wider considerations of political economy, in particular to the protection of employment, regional development and the trade balance. This policy was reversed from 1977, culminating in privatization, 'the apotheosis of insularity' [Wilks, 1990, *176*].

Wilks's emphasis on the insular nature of relationships and Reich's stress on the weakening effect of the non-discriminatory policies of successive British governments imbued with the ethos of a traditional liberal, market ideology need to be put into an international context. Whereas Dunnett was critical of government because intervention undermined the competitive process, which he implies would have produced a creative response from the British industry, the criticism made by Wilks and Reich focused on the form which the intervention took, comparing the role of government in other countries involving the selection and systematic support of core firms or of a national champion.

Historians are not agreed on the extent of the contrast between government–industry relations in Britain and on the Continent, the extreme case most often cited being that of Germany. Abromeit [1990] stressed the liberal market ideology of successive post-war German governments, pointing out that contacts and communications between government and industry were good and that they focused mainly on questions concerning competitive structure and environmental regulation policies. He rejects the view that government support for industry was the product of a coherent industrial policy, let alone a strategy of industrial targeting or sectoral modernization. Measures were *ad hoc*, including fiscal measures to affect the climate for investment, the setting up of advisory bodies and the financing of research and development, and regional finance for restructuring declining industries.

Where government provided direct support, for Daimler and Benz for example, Abromeit emphasized that initiatives almost invariably originated from individual companies rather than from government. This, he argued, was evidence not of the power of trade associations but of the economic strength of large business organizations, whose influence owed more to the direct involvement of banks than to government–industry relations and discriminating government policies. Abromeit's account of successive German governments' attempts to avoid involvement and of intervention signalling 'crisis management', bears a distinct resemblance to government relations with the British motor industry.

Because British companies had remained insular their inability to cushion the cumulative effects of free trade and internationalization was in marked contrast to the options available to the multinationals. Tied imports and sourcing offered them some escape from the labour and production difficulties which also affected the industry in Britain. Such a strategy enabled Ford and Vauxhall to succeed in expanding their share of the British market, where prices and profit margins were high relative to the rest of Europe [Ashworth, Kay and Sharpe, 1982]. None the less, Ford UK's inferior record compared with its European counterparts is evidence that it too experienced some of the debilitating effects of being part of a defective national economy. It thereby experienced government policies which were not only destabilizing to investment and indirectly contributed to the militancy of car workers, but also produced slower national economic growth than that experienced by other European countries, where real incomes were higher [Jones, 1981]. Together these circumstances presented an economic, social, and political environment less conducive to the achievement of the highest European standards of manufacturing efficiency. In this respect the history of the decline of the motor industry is of wider significance for an understanding of the record of British manufacturing industry.

There are, however, certain factors the importance of which varied over time yet which were specific in contributing to the ultimate termination of Britain's only major independent motor manufacturer. Throughout the post-war period, but especially during the crucial years between the mid 1950s and the mid 1960s, BMC was producing models which were in demand at home and overseas, productivity levels were comparable still with those of manufacturers on the Continent. Corporate weakness, however, inhibited a policy of exploiting strength in the market, introducing structural and organizational changes affecting production and marketing, accompanied by investment in model development.

The particular difficulties presented by government policies and industrial relations to all motor manufacturers in Britain are not in doubt. Yet the rapid decline of the British motor industry during the 1970s was in part the outcome of repeated failures by managers to distinguish between short- and long-term problems. During the post-war boom which lasted until 1965 growth had concealed the industry's weaknesses as reported by the NAC and PEP in 1945 and 1950. Short-term instability in demand had been met by temporary reductions in capacity, partly by laying workers off, partly by allowing stoppages to drag on during recessions.

Thereafter, the years of static home demand were also attributed within the industry to government policies: the dip in exports to an overvalued currency which devaluation in 1967 appeared to have corrected when exports surged until 1970. This apparent failure on the part of British managers to distinguish short-term difficulties and their local causes from longer term trends was exacerbated by a lack of strategic planning. Furthermore, the investment that did take place during the 1960s and early 1970s was undertaken without vital information on production costs. This critical gap, crucial for designing cars to make profits for reinvestment, also precluded comparisons between the profitability of different models. Lack of information seriously hampered the formulation of informed production and marketing strategies. Simultaneously, and in part as a consequence, the inability to invest in new models led to a deterioration in the company's engineering and design capacity [Whipp and Clark, 1986]. Lack of information, weakness of corporate integration and inadequate managerial resources were a combination which proved to be ill-suited to enable BMC, BMH or BLMC either to restructure themselves effectively or to adapt successfully to the rapidly changing internationally competitive economic environment.

These features were the legacy of corporate cultures established well before the Second World War, creating the limits within which managers exercised what economists have called 'bounded rationality', a concept which allows for rational intentions leading to unintended and less than optimal outcomes. The autocratic managerial approach of the founders of those companies which survived the competition of the interwar years was followed by a personal style adopted by successor managers (who were not family inheritors) from within those organizations. Even after the Leyland merger in 1968 the managerial style perpetuated was that characteristic of a personal enterprise in an industry and in an era which required a transition to some form of professional, managerial organization and control. In a world dominated by international and multinational business international comparisons were central to corporate decisions on which, how, and where cars should be produced. The British companies lacked the organizational capacity to assess systematically their competitive environment and to respond. By contributing to a climate of uncertainty both government and labour added to the industry's problems, one effect of which was to discourage long-term planning. Yet without the introduction of accurate costing, managers were in no position to recognize the point when their organization reached the critical level of vulnerability beyond which, unless fundamental changes in organization were made, a downward spiral would inevitably ensue.

The failure of the British motor industry to transform itself into an internationally competitive enterprise is explicable therefore mainly by three interacting factors. First were government policies in which political considerations constrained business decision-making and assisted the multinationals' strength in the domestic market. Second there was a system of industrial relations which was rooted neither in law nor in trade union power. Third, and fundamentally, there were historically rooted weaknesses in corporate structures and management which for so many years obscured the need for systematic planning and organizational change. When government finally took command during the late 1970s, the task of rescuing BLMC/BL had become more problematical as a result of simultaneous developments in the dynamics of the industry, notably the trend towards globalization and changes in the character of demand and competition in the domestic as well as in international markets. Neither nationalization nor privatization prevented Britain from becoming the first major car-producing country to relinquish a domestically owned independent national champion.

Bibliography

The place of publication is London unless otherwise indicated.

Abromeit, H. (1990) 'Government–industry Relations in West Germany', in Chick, M. (ed.), *Governments, Industries and Markets* (Aldershot).

Adeney, Martin (1989) *The Motor Makers.*

Alford, B. W. E. (1972) *Depression and Recovery? British Economic Growth.*

Alford, B. W. E. (1981) 'New industries for old? British industry between the wars' in Floud, R. and McCloskey, D. (eds), *The Economic History of Britain since 1700*, Vol. 2, *1860 to the 1970s* (Cambridge).

Alford, B. W. E. (1986) 'Lost opportunities: British business and businessmen during the First World War', in McKendrick, N. and Outhwaite, R. B. (eds), *Business Life and Public Policy* (Cambridge).

Alford, B. W. E. (1988) *British Economic Performance, 1945–1975* (Basingstoke).

Allen, G. C. (1926) 'The British Motor Industry', *London and Cambridge Economics Service*, LSE Special Memo, No. 18.

Andrews, P. W. S. and Brunner, E. (1955) *The Life of Lord Nuffield* (Oxford).

Armstrong, A. G. (1967) 'The motor industry and the British economy', *District Bank Review.*

Ashworth, M. H., Kay, J. A., and Sharpe, T. A. E. (1982) *Differentials between car prices in the United Kingdom and Belgium*, IFS Report, Series no. 2 (Institute for Fiscal Studies).

Bardou, J.-P., Chanaron, J. J., Fridenson, P., and Laux, J. M. (1982) *The Automobile Revolution: The Impact of an Industry.* [Translated from French by J. M. Laux] (Chapel Hill, North Carolina, US).

Barker, T. C. (1982) 'The Spread of Motor Vehicles before 1914' in Kindleberger and di Tella (eds), *Economics in the Long View*, Vol. 2.

Barnett, C. (1986) *The Audit of War: the Illusion and Reality of Britain as a Great Nation.*

Benn, Tony (1988) *Office Without Power, Diaries 1968–1972.*

Benn, Tony (1989) *Against the Tide, Diaries, 1973–76.*

Bevan, D. L. (1977) 'The Nationalized Industries', *The Economic System in the UK,* ed. Morris, D. (Oxford).

Beynon, H. *Working for Ford* (Harmondsworth, 1973).

Bhaskar, K. (1975) *Alternatives Open to the UK Motor Vehicle Industry* (Bath)

Bhaskar, K. (1979) *The Future of the UK Motor Industry.*

Bhaskar, K. (1983) *The Future of the UK and European Motor Industry* (Bath).

Bhaskar, K. *et al.* (1984a) *Research Report on the Future of the UK and the European Motor Industry,* Vols I and II (Bath).

Bhaskar, K. *et al.* (1984b) *State Aid to the European Motor Industry: a Report* (Norwich).

Blaich, F. (1981) 'The development of the distribution sector in the German car industry' in Okochi, Akio, Shimokawa, and Koichi (eds), *Development of Mass Marketing* (Tokyo).

Blaich, F. (1987) 'Why did the pioneer fall behind? Motorization in Germany between the wars' in Barker, T. (ed.), *The Economic and Social Effects of the Spread of Motor Vehicles.*

Bloomfield, G. (1978) *The World Automotive Industry.*

Bowden, S. M. (1991) 'Demand and supply constraints in the interwar car industry. Did the manufacturers get it right?' *Business History,* 33, 2.

Cannell, R. *et al.* (1984) *Ford of Europe: A Strategic Profile.*

Capie, F. (1983) *Depression and Protectionism: Britain between the Wars.*

Carr, Christopher (1990) *Britain's Competitiveness. The Management of the Motor Vehicle Components Industry.*

Carr, F. W. (1978) 'Engineering workers and the rise of labour in Coventry, 1914–1939' (Warwick, PhD thesis).

Castle, Barbara (1980) *The Castle Diaries.*

Caunter, C. F. (1957) *The History of Development of Light Cars,* HMSO.

Caunter, C. F. (1970) *The Light Car, a Technical History.*

Central Policy Review Staff (1975) *The Future of the British Car Industry,* HMSO.

Chandler, A. D. (1990) *Scale and Scope: The Dynamics of Industrial Capitalism.*

Channon, D. (1973) *The Strategy and Structure of British Enterprise* (Boston).

Chapman, A. L. and Knight, R. (1953) *Wages and Salaries in the United Kingdom 1920–38* (Cambridge).

Church, R. A. and Miller, M. (1977) 'The Big Three' in Barry Supple (ed.), *Essays in British Business History* (Oxford).

Church, R. A. (1977) 'Myths, Men and Motorcars', *Journal of Transport History*, IV, No. 2.

Church, R. A. (1978) 'Innovation, monopoly and supply of vehicle components in Britain, 1880–1930: the growth of Joseph Lucas Ltd', *Business History Review*, LII.

Church, R. A. (1979) *Herbert Austin: The British Motor Car Industry to 1941*.

Church, R. A. (1981) 'The Marketing of Automobiles in Britain and the United States before 1939', Okochi A. and Koichi S. (eds), *The Development of Mass Marketing* (Tokyo).

Church, R. A. (1982) 'Markets and Marketing in the British Motor Industry before 1914', *Journal of Transport History*, Spring 1982.

Church, R. A. (1986a) 'The Effect of American Multinationals on the British Motor Industry' in Levy Leboyer, M. and Teichova, A. (eds), *Multinationals in Historical Perspective* (Cambridge).

Church, R. A. (1986b) 'Family firms and managerial capitalism: the case of the international motor industry', *Business History*, 28, 2, 165–80.

Church, R. A. and Mullen, C. (1989) 'Cars and corporate culture: the view from Longbridge' in B. Tilson (ed.), *Made in Birmingham: Design and Industry, 1889–1989* (Brewin, Studley, Worcs).

Church, R. A. (1993) 'The mass marketing of motor cars before 1950: The missing dimension' in Tedlow, R. and Jones, G. (eds), *The Rise and Fall of Mass Marketing*.

Clack, G. (1967) *Industrial Relations in a British Car Factory* (Cambridge).

Clayden, Tim (1987) 'Trade unions, employers and industrial relations in the British motor industry', *Business History*, 29, 3, 304–24.

Cole, G. D. H. (1923) *Workshop Organisation*.

Collins, M. (1991) *Banks and Industrial Finance in Britain, 1830–1939*.

Coppock, D. J. (1956) 'The climacteric of the 1890s: a critical note', *Manchester School*, XXIV.

Cottrell, P. L. (1980) *Industrial Finance, 1830–1914. The Finance and Organisation of English Manufacturing Industry*.

Cowling, M. (1980) 'The Motor Industry' in idem, *Mergers and Economic Performance*.

Cowling, M. (1986) 'The Internationalization of Production and Deindustrialization', in Amin, A. and Goddard, J. (eds), *Technological Change, Industrial Restructuring and Regional Development*.

Croucher, R. (1982) *Engineers at War*.

Davenport-Hines, R. P. T. (1984) *Dudley Docker, the Life and Times of a Trade Warrior* (Cambridge).

Davy, J. (1967) *The Standard Car 1903–1963* (Coventry).

Department of Industry (1976) cmnd 6377 *The British Motor Vehicle Industry*.

Donnelly, T. and Thoms, D. (1990) 'Trade Unions, Management and the Search for Production in the Coventry Motor Car Industry, 1939–75', *Business History*, 31.

Donoghue, B. (1987) *Prime Minister: The Conduct of Policy under Harold Wilson and James Callaghan*.

Dunnett, P. J. S. (1980) *The Decline of the British Motor Industry*.

Durcan, J. W., McCarthy, W. E. J., and Redman, G. P. (1983) *Strikes in Post-war Britain: A study of stoppages of work due to industrial disputes, 1946–1973*.

Edelstein, M. (1976) 'Realized rates of return in UK home and overseas portfolio investment in the age of high imperialism', *Explorations in Economic History*, 13, 283–329.

Edelstein, M. (1982) *Overseas Investment in the Age of High Imperialism in the United Kingdom, 1850–1914*.

Edwardes, Michael (1983) *Back from the Brink*.

Edwards, P. (1982) 'Britain's changing strike problem', *Industrial Relations Journal*, 13, no. 2.

Elbaum, B. and Lazonick, W. (1986) 'An institutional perspective on British decline', *The Decline of the British Economy*, ed. Elbaum, B. and Lazonick, W. (Oxford).

Engelbach, C. R. F. (1927–28) 'Some notes on re-organizing a works to increase production', *Proceedings of the Institute of Automobile Engineers*, XXII.

Engelbach, C. R. F. (1933–4) 'Problems in manufacture', *Proceedings of the Institute of Automobile Engineers*, XXVIII.

Flanders, A. (1952) 'Industrial relations' in Worswick, G. N. and Ady, P. H. (eds), *The British Economy, 1945–50* (Oxford).

Foreman-Peck, J. (1979) 'Tariff protection and economies of scale: the British motor industry before 1939', *Oxford Economic Papers*, 31, No. 2.

Foreman-Peck, J. (1981a) 'The effect of market failure on the British motor industry before 1939', *Explorations in Economic History*, 18.

Foreman-Peck, J. (1981b) 'Exit, voice and loyalty as responses to decline: the Rover Company in the interwar years', *Business History*, 23, No. 2.

Foreman-Peck, J. (1982) 'The American challenge of the Twenties and the European motor industry', *Journal of Economic History*, XLII, No. 4.

Foreman-Peck, J. (1983) 'Diversification and the growth of the firm: the Rover company to 1914', *Business History*, 25, No. 2.

Foreman-Peck, J. (1985) 'Intra-firm trade in the international motor vehicle industry', in Casson, M. (ed.), *Multinational Companies and World Trade*.

Fridenson, P. (1978) 'The Coming of the Assembly Line to Europe' in W. Krohn (ed.), *The Dynamics of Science and Technology*, Vol. II (Dordrecht).

Friedman, A. L. (1977) *Industry and Labour: Class Struggle at Work and Monopoly Capitalism*.

Friedman, A. L. (1984) 'Management strategies, market conditions and the labour process', in F. J. Stephens (ed.), *Firms Organization and Labour*.

Friedman, H. and Meredeen, S. (1980) *The Dynamics of Industrial Conflict: Lessons from Ford*.

Frostick, Michael (1970) *Advertising the Motor Car*.

Gennard, J. and Steuer, M. D. (1971) 'The Industrial Relations of Foreign-owned Subsidiaries in the United Kingdom', *British Journal of Industrial Relations*, IX, 1.

Griffith, F. (1955) 'Why Austin Developed Unit Construction Transfer Machines', *The Machinist*, January.

Hackett, Dennis (1978) *The Big Idea. The Story of Ford in Europe* (Ford Motor Co.).

Hague, D. C. and Williamson, G. (1983) *The I. R. C. An Experiment in Industrial Intervention*.

Hannah, L. (ed.) (1976) *Management Strategy and Business Development*.

Hannah, L. (1983) *The Rise of the Corporate Economy*.

Harrison, A. E. (1981) 'Joint Stock flotation in the cycle, motor cycle and related industries, 1882–1914', *Business History*, 23.

Harrison, A. E. (1982) 'F. Hopper and Co: The Problem of Capital Supply in the Cycle Manufacturing Industry, 1891–1914', *Business History*, 24, 3–23.

Hinton, James (1973) *The First Shop Stewards' Movement*.

Holden, L. T. (1984) 'A History of Vauxhall Motors to 1950: Industry, Development and Local Impact on the Luton Economy' (M. Phil., Open University).

Hope, A. (1979) 'The Genius Today', *Autocar*, 25 August.

Hornby, William (1958) *Factories and Plant*.

Hounshell, D. A. (1984) *From the American System to Mass Production 1800–1932* (Baltimore).

House of Commons (1975) *British Leyland: the next decade* (The Ryder Report). HC 342, 1974/5.

House of Commons (1975) *The Motor Vehicle Industry*, 14th Report of the Trade and Industry Sub-Committee of the Expenditure Committee, HC 617.

House of Commons (1976) *Chrysler UK*, 8th Report of the Trade and Industry Sub-Committee of the Expenditure Committee, HC 146, 1975–6.

House of Commons (1977) *British Leyland*, Minutes of Evidence taken before the Expenditure Committee, HC 396, 1976–7.

House of Commons (1987) *The UK Components Industry*, Third Report of the Trade and Industry Select Committee, HC 407, 1986–7.

Hyman, Richard (1971) *The Workers' Union* (Oxford).

Imperial Economic Committee, Thirteenth Report (1936) *A Survey of the Trade in Motor Vehicles*.

Irving, R. J. (1975) 'New industries for old: some investment decisions of Armstrong Whitworth, 1900–1914', *Business History*, 17.

Jeffreys, J. B. (1945) *The Story of the Engineers*.

Jeremy, David (ed.) (1984–6) *A Biographical Dictionary of Business Leaders Active in Britain in the Period 1860–1980*.

Jones, D. T. (1981) *Maturity and Crisis in the European Car Industry*, Sussex European Papers No. 8, Brighton.

Jones, D. T. (1983) 'Technology and the UK Automobile Industry', *Lloyds Bank Review*, No. 148, April, 14–27.

Jones, D. T. (1984) 'Technology and the UK Automobile Industry', *Lloyds Bank Review*, April.

Jones, D. T. (1985) *The Import Threat to the UK Car Industry*, Science Policy Research Unit, Brighton.

Jones, D. T. and Prais, S. J. (1978) 'Plant Size and Productivity in the Motor Industry: Some International Comparisons', *Oxford Bulletin of Economics and Statistics*, 40, No. 2, May.

Kennedy, W. P. (1976) 'Institutional Response to Economic Growth: Capital Markets in Britain to 1914' in L. Hannah (ed.), *Management Strategy and Business Development: An Historical and Comparative Study*, pp. 151–83.

Khan, A. E. (1946) *Great Britain and the World Economy.*

Lambert, Z. E. and R. J. Wyatt (1968) *Lord Austin, the Man.*

Laux, James (1976) *In First Gear* (Liverpool).

Law, C. M. (1985) 'The Geography of Industrial Rationalisation: The British Motor Car Industry, 1972–1982', *Geography*, 40, January.

Lewchuck, W. A. (1983) 'Fordism and the British Motor Car Employers, 1896–1932' in H. Gospel and G. Littler (eds), *Managerial Strategies and Industrial Relations.*

Lewchuck, W. A. (1984) 'The role of the British government in the spread of scientific management and Fordism in the interwar years', *Journal of Economic History*, 44, 355–61.

Lewchuck, W. A. (1985a) 'The Return to Capital in the British Motor Vehicle Industry', *Business History*, 27.

Lewchuck, W. A. (1985b) 'The origins of Fordism and alternative strategies: Britain and the United States, 1880–1930' in Tolliday, S. and Zeitlin, J. (eds), *Between Fordism and Flexibility, the International Motor Industry and its Workers* (Oxford).

Lewchuck, W. A. (1986) 'The Motor Vehicle Industry' in Elbaum, B. and Lazonick, W. (eds), *The Decline of the British Economy* (Oxford).

Lewchuck, W. A. (1987) *American Technology and the British Motor Vehicle Industry* (Cambridge).

Lloyd, I. (1978) *Rolls Royce, the Growth of a Firm.*

Lloyd, I. (1978) *Rolls Royce, Years of Endeavour.*

Locke, R. R. (1984) *The End of Practical Man: Entrepreneurship and Higher Education in Germany, France and Great Britain, 1880–1940*, Vol. VIII ed. McKay, J. P. (Greenwich, Conn.).

Lyddon, D. (1983) 'Workplace organization in the British car industry', *History Workshop*, 15.

Mackay, D. J., Sladen, Janet P., and Halligan, Margaret J. (1984) *The UK Vehicle Manufacturing Industry: Its Economic Significance* (PEIDA).

Marsden, David, Morris, Timothy, Willman, Paul, and Wood, Stephen (1985) *The Car Industry: Labour Relations and Industrial Adjustment.*

Mathias, P. (1983) second edn, *The First Industrial Nation.*

Maxcy, G. and Silberston, A. (1959) *The British Motor Industry.*

Maxcy, G. (1981) *The Multinational Motor Industry.*

Maxcy, G. (1958) 'The Motor Industry' in Cook, P. L. and Cohen, R. (eds), *Effects of Mergers.*

Melman, S. (1958) *Decision Making and Productivity* (Oxford).

Michie, R. C. (1981) 'Options, Concessions, Syndicates and the Provision of Venture Capital, 1880–1913', *Business History*, 23, 147–64.

Miller, M. and Church, R. A. (1979) 'Growth and Instability in the British Motor Industry between the Wars' in Aldcroft, D. H. and Buxton, C. *Instability and Industrial Development, 1919–1939.*

Ministry of Supply (1947) *National Advisory Council for the Motor Manufacturing Industry: Report of Proceedings* (HMSO).

Mitchell, B. R. (1962) *Abstract of European Historical Statistics.*

Morris, W. R. (1924) 'Policies that have built the Morris Motor Business', *System*, February.

Musson, A. E. (1978) *The Growth of British Industry.*

National Advisory Council, Ministry of Supply (1947) *Report and Proceedings Council, Motor Manufacturing Industry*, HMSO.

National Economic Development Council (1970) *Industrial Report by the Motor Manufacturing Development Council on the Economic Assessment to 1972.*

Nelson, W. H. (1967) *Small Wonder, the Amazing Story of the Volkswagen.*

Nicholson, T. R. (1983) *The Birth of the British Car Industry*, Vols I, II, III.

Nockolds, H. (1976) *Lucas, the First Hundred Years*, Vol. I.

Nubel, Otto (1987) 'The Beginnings of the Automobile in Germany' in Barker, T. (ed.), *The Economic and Social Effects of the Spread of Motor Vehicles.*

OECD (1987) *The Cost of Restricting Imports: The Automobile Industry.*

Oliver, George (1971) *The Rover.*

Overy, R. J. (1973) 'Transportation and Rearmament in the Third Reich', *Historical Journal*, 16, 390–411.

Overy, R. J. (1975) 'Cars, roads and economic recovery' *Economic History Review*, XXVIII, 3, 466–83.

Overy, R. J. (1976) *William Morris, Viscount Nuffield.*

Pagnamenta, P. and Overy, R. J. (1984) *All Our Working Lives* (BBC, London).

Payne, P. L. (1988) *British Entrepreneurship in the Nineteenth Century* (second edn).

Pettigrew, A. (1979) 'On studying organisational cultures', *Administrative Science Quarterly*, 24, 4.

Plowden, W. E. (1971) *The Motor Car and Politics*.

Political and Economic Planning (PEP) (1950) Motor Vehicles, Engineering Report II.

Pollard, Sidney (1982) *The Wasting of the British Economy*.

Porter, M. (1989) *Competitiveness in the International Economy* (New York).

Pratten, C. and Silberston, A. (1967) 'International comparisons of labour productivity in the automobile industry, 1955–65', *Bulletin of the Oxford Institute of Statistics*, August 1967.

Pratten, C. (1971) *Economies of Scale in British Manufacturing Industry*, Cambridge.

Pryke, R. (1981) *The Nationalised Industries, Policies and Performance since 1968* (Oxford).

Rae, J. B. (1959) *American Automobile Manufacturers. The First Forty Years* (Philadelphia).

Reich, S. (1990) *Fruits of Fascism: Post-war Prosperity in Historical Perspective* (Cornell).

Report of the Liberal Industrial Inquiry (1928): Britain's Industrial Future.

Rhys, D. G. (1972) *The Motor Industry: An Economic Survey*.

Rhys, D. G. (1974) 'Employment Efficiency and Labour Relations in the British Motor Industry', *Industrial Relations Journal*, 5, No. 2.

Rhys, D. G. (1976) 'Concentration in the Inter-war Motor Industry', *Journal of Transport History*, New Series III, 4, 241–64.

Rhys, D. G. (1977) 'European mass-producing car makers and minimum efficient scale', *Journal of Industrial Economics*, XXV.

Rhys, D. G. (1988) 'Motor Vehicles' in Johnson, P. (ed.), *The Structure of British Industry*.

Richardson, H. W. (1961) 'The new industries between the wars', *Oxford Economic Papers*, XIII.

Richardson, H. W. (1965) 'Over-commitment in Britain before 1930', *Oxford Economic Papers*, XVII.

Richardson, Kenneth (1972) *Twentieth Century Coventry* (Coventry).

Rostas, L. (1948) *Corporative Productivity in British and American Industry* (Cambridge).

Rostow, W. W. (1963) *The Economics of Take-off into Sustained Growth.*

Royal Commission on Trade Unions and Employer Associations (1968) (Donavon Report), Cmd. 1623.

Salmon, E. A. (1975) 'Inside BL', *Management Today*, November.

Samuels, R. J. (1990) 'Business and the Japanese state', in Chick and Martin (eds), *Governments, Industries and Markets*, p. 37.

Sargant Florence, P. (1953) *Logic of British and American Industry.*

Sargant Florence, P. (1961) *Ownership, Control and Success of Large Companies.*

Saul, S. B. (1962) 'The Motor Industry in Britain to 1914', *Business History*, 5.

Saul, S. B. (1968) 'The Engineering Industry' in Aldcroft, D. H., *The Development of British Industry.*

Sedgwick, M. (1970) *Cars of the 1930s.*

Sedgwick, M. (1975) *Passenger Cars, 1924–1942.*

Silberston, A. (1958) 'The Motor Industry' in Burn, D. L. (ed.), *The Structure of British Industry* (Cambridge).

Silberston, A. (1965) 'The Motor Industry 1955–1964', *Oxford Bulletin of Economics and Statistics*, Vol. 27.

Skilleter, Paul (1988) 'The Thomas Papers', *Thoroughbred and Classic Cars*, June.

Sloan, Alfred, P., Jr (1967) *My Life with General Motors.*

SMMT, Society of Motor Manufacturers and Traders.

Starkey, Ken and McKinlay, Alan (1988) *Organizational Innovation, Competitive Strategy and the Management of Change in Four Major Companies* (Avebury).

Streeck, W. (1984) *Industrial Relations in West Germany.*

Sugden, Roger, 'The Warm Welcome for Foreign-owned Transnationals from recent British Governments', in Chick, M. (ed.), *Governments, Industries and Markets* (Aldershot).

Thomas, Sir Miles (1964) *Out on a Wing.*

Thomas, R. P. (1973) 'Style change and the automobile industry during the roaring twenties' in Cain, L. P. and Uselding, P., *Business Enterprise and Economic Change* (Chicago).

Thoms, David and Donnelly, Tom (1985) *The Motor Car Industry in Coventry since the 1890s.*

Thornhill, A. R. (1986) 'Industrial Relations in the British Motor Industry to 1939' (PhD, University of East Anglia).

Tiratsoo, N. (1992), 'The Motor Car Industry', in Mercer, H., Rollings, N., and Tomlinson, J. D., *Labour Governments and Private Industry* (Edinburgh).

Tolliday, S. (1983) 'Trade unions and shop floor bargaining in the British motor industry, 1910–1939', *Bulletin of the Society for the Study of Labour History,* 46, Spring.

Tolliday, S. (1985) 'Government, Employers and Shop Floor Organisation in the British Motor Car Industry 1939–1969' in Tolliday, S. and Zeitlin, J. (eds), *Shop Floor Bargaining and the State* (Cambridge).

Tolliday, S. (1986) 'High tide and after: Coventry engineering workers and shop floor bargaining 1945–80' in Lancaster, A. and Mason, T. (eds), *Life and Labour in a Twentieth Century City: the Experience of Coventry* (Coventry).

Tolliday, S. (1987a) 'The failure of mass production unionism in the motor industry, 1914–1939' in Wrigley, C. (ed.), *A History of British Industrial Relations,* Vol. 2, *1914–1939.*

Tolliday, S. (1987b) 'Management and Labour, 1896–1939' in Tolliday, S. and Zeitlin, J. (eds), *The Automobile Industry and its Workers: Between Fordism and Flexibility* (Cambridge).

Tolliday, S. (1991) 'Ford and "Fordism" in postwar Britain: enterprise management and the control of labour, 1937–1987' in Tolliday, S. and Zeitlin, J. (eds), *The Power to Manage.*

Turner, Graham (1971) *The Leyland Papers* (Birkenhead).

Turner, H. A., Clack, G. and Roberts, G. (1967) *Labour Relations in the Motor Industry.*

Tweedale, G. (1987) 'Business and investment strategies in the interwar British steel industry: case study of Hadfields Ltd. and Bean Cars', *Business History,* 29.

US Bureau of the Census (1976) *The Statistical History of the United States from Colonial Times to the Present* (New York).

US Bureau of Domestic and Foreign Commerce (1928) *Automotive Industry and Trade of Great Britain* (New York).

Von Tunzelman, G. N. (1978) 'Structural change and leading sectors in British manufacturing industry, 1907–68' in Kindleberger, C. P. and di Tella (eds), *Economics in the Long View.*

Waymark, P. (1983) *The Car Industry.*

Wells, L. T. (1974) 'Automobiles' in Vernon, R. (ed.), *Big Business and the State: Changing Relations in Western Europe.*

Whipp, R. (1987) 'Technology, Management, Strategic Change and Competitiveness', in Dorgham, M. A. (ed.), *Proceedings of the Fourth International Vehicle Design Congress* (Geneva).

Whipp, R. and Clark, P. (1986) *Innovation and the Automobile Industry, Product Process and Work Organization* (New York).

Whipp, R., Rosenfeld, R., and Pettigrew, A. (1987) 'Understanding Strategic Change Processes: Some Preliminary Findings' in Pettigrew, A. (ed.), *The Management of Strategic Change* (Oxford).

Whiting, R. C. (1983) *The View from Cowley: The Impact of Industrialization upon Oxford, 1918–38* (Oxford).

Wigham, E. (1973) *The Power to Manage.*

Wild, R. (1974) 'The origins and development of flow-line production', *Industrial Archaeology*, II.

Wilkins, Mira and Hill, Frank Ernest (1964) *American Business Abroad: Ford on Six Continents* (Detroit).

Wilks, S. (1984) *Industrial Policy and the Motor Industry* (Manchester).

Wilks, S. (1990) 'Institutional Insularity: Government and the British Motor Industry Since 1945', in Chick, M. (ed.), *Governments, Industries and Markets* (Aldershot).

Williams, K. J., Williams, J. and Thomas, Dennis (1983) *Why are the British Bad at Manufacturing?*

Williams, K., Williams, J. and Haslam, C. (1987) *The Breakdown of Austin Rover* (Leamington Spa).

Willman, P. (1984) 'The Reform of Collective Bargaining and Strikes at BL Cars', *Industrial Relations Journal*, 15, No. 2.

Willman, P. (1987) 'Labour relations strategies at BL cars' in *The International Automobile Industry and its Workers: Between Fordism and Flexibility*, ed. Tolliday, S. and Zeitlin, J. Z. (Oxford).

Willman, P. (1986) *Technological Change, Collective Bargaining and Industrial Efficiency* (Oxford).

Willman, P. and Winch, G. (1987) *Innovation and Management Control, Labour Relations at BL Cars* (Cambridge).

Wilson, Harold (1971) *The Labour Government, 1964–1970: A Personal Record.*

Wilson, Harold (1979) *The Final Term: The Labour Government, 1974–6.*

Wood, Jonathan (1988) *Wheels of Misfortune.*

Woollard, F. W. (1954) *The Principles of Flow Production.*

Wyatt, R. J. (1968) *The Motor for the Million: The Austin Seven 1922–1939.*

Wyatt, R. J. (1981) *The Austin, 1905–52.*

Young, S. and Hood, N. (1977) *Chrysler UK: A Corporation in Transition* (New York).

Zeitlin, J. (1979) 'Craft control and the division of labour: engineers and compositors in Britain, 1890–1930', *Cambridge Journal of Economics*, 3, 263–74.

Zeitlin, J. (1980) 'The Emergence of Shop Steward Organisation and Job Control in the British Car Industry', *History Workshop*, 10.

Zeitlin, J. (1983) 'Workplace Militancy: A Rejoinder', *History Workshop*, 16.

Zeitlin, J. (1983) 'The Labour Strategies of British Engineering Employers, 1890–1922' in *Managerial Strategies and Industrial relations*, ed. Gospel, H. and Littler, C.

Index

Amalgamated Engineering
  Union 43, 63, 64
American industry 3, 18, 32, 49, 65
asset growth 27, 55, 72, 78
Austin, Herbert 1, 30, 31, 33, 78
Austin Motor Co. 8, 9, 23, 25, 27–8,
  30, 35, 40, 75–7, 79, 80–1

balance of trade 44, 56, 86, 113
banks 40, 99
Benn, Anthony Wedgwood 85, 90,
  100
British Aerospace 107, 114–15
British Leyland Motor Corpora-
  tion/BLMC/BL 49, 76, 79,
  84, 86–9, 90–106, 108, 112,
  114, 118
British Motor Corporation
  (BMC) 64, 72, 78–80, 82, 91,
  98, 115, 121–2
British Motor Holdings 49, 79, 83,
  85–6, 88, 93

capital, investment 55, 57–8, 71–2,
  99, 112
  return on 39, 61
Central Policy Review Staff 87
Chrysler 30, 85, 90, 108
  UK 30, 49, 79, 113, 119, 120
Citröen 37, 108
commercial vehicles 16, 44, 84,
  90–1, 106
competition, non-price 32–3, 38, 96
  price 36–7, 38
components, assembly 39, 103–4
  industry 113
corporate culture 102, 124

costs, motoring 18
  production 48, 81
Coventry and District Engineering
  Association 43
Coventry tool room agreement 43

dealers 57, 93, 98
demand 11–13, 44
  instability 56–7, 59
design 2, 16, 33–4, 55, 79, 92, 103
distribution 57
dividends 26–7, 28–9, 72, 99

economic growth 46, 50, 51
economies of scale 61, 75, 76, 92–3,
  95, 104
Edwardes, Michael 101, 102, 106
Emergency Powers Act 43
employment 51, 107
engineering 1–3
Engineering Employers'
  Federation 8, 22–3, 25, 62–3,
  67
entrepreneurs 1–3
European Economic
  Community 47, 96
exports 19, 20, 44, 51, 56–7, 97–8,
  104, 111–23

Fiat 108
finance 1, 16, 27, 71, 74
Ford, Henry 59
Ford, model T 5–6, 15, 34
Ford Motor Company 5, 32, 74,
  108
  of Europe 73, 88, 108, 109
  in Germany 57, 74, 109, 119

in the UK 5–6, 8–9, 16, 21, 26–7, 38, 40, 54–8, 62, 69–72, 74, 76, 78–9, 82, 94, 109–13, 119, 122
Fordism 21, 32, 34
French industry 1, 3–4, 15, 41, 44, 46, 51, 58

General Motors 32, 54, 59, 108–9, 111
German industry 1, 3, 15, 41, 44, 46, 48–9, 51, 53, 58, 65, 72, 76–7, 96, 117
globalization 108–9
government, relations with industry 43, 50–3, 56–7, 59, 73, 85, 100, 118–19, 120–1
in Germany 121–2
subsidies 59, 100, 107, 112–13

Hillman Motor Company 7, 30, 40

imports 10, 16, 96, 99, 108
tied 109–10
Industrial Reorganization Corporation 85, 116, 118
Italian industry 46, 54

Jaguar Motor Co. 75, 77, 79, 88, 94, 106
Japanese industry 96, 108, 114–15

labour 8, 23–5, 68
management 22, 61, 115
relations 60, 69, 101–2, 116–18
Lancashire Steam Motor Co. 90
Leyland Motors 49, 84, 90
Lord, L. P. 62, 80

management 80, 88–91, 100–1, 106–7, 116, 118, 124
marketing 33, 37–8, 53, 80–1, 98, 104–5
markets 5–6, 14–15, 32, 95
approach to 2, 5, 34–6, 53, 94
shares 34, 36, 95
mass production, 'British system' of see payment systems, production methods

mergers 40, 49, 78–84, 88–92, 115–16, 118
Mini 80, 82–3, 92
Morris Motors 7, 10, 18, 23–4, 26–8, 34, 36–7, 79, 81
Morris, W. R. (Lord Nuffield) 6, 8, 13, 28, 31, 35, 40, 53, 78
multinationals 107–8, 111, 122
munitions 8–9
'mutuality' 63

National Advisory Council 46, 52
National Enterprise Board 100–1
National Union of Vehicle Builders 24
Nissan 79, 114
Nuffield organization 75, 80, see Morris

output see production
overmanning 49, 87, 93, 114, 116
owner-managed firms 26, 30–1, 90
ownership, car 2, 17–18, 44–5

payment systems 22–3, 62–4, 67, 69–70
Political Economic Planning 47, 75
Perry, Percival 5, 9, 15, 73, 74
Peugeot 5, 79, 108
piece rate see payment systems
plant size 61, 66, 76
prices 12–13, 19, 81
privatization 106–7
production 2, 18, 44–5, 55, 79, 107–8
production methods 3–4, 6, 21–2, 24–6, 35, 61, 105
productivity 48–9, 57–8, 65–8, 96, 105, 114
profits/losses 27–8, 40, 72, 82, 93–4, 99, 106, 112

quality 98, 114

Renault 58, 108
research and development 111
Rolls Royce 75, 77, 79
Rootes Group 30, 38–40, 49, 72, 75, 79, 85

Rootes, Lord 90, 92
Rover Motor Co. 29, 40, 75, 77, 79,
    84, 88–9, 94–5, 107, 114–15
Ryder Report 87, 100, 102

shadow factories 43
shop stewards 22, 24, 62–4, 70–1,
    116
Singer Motor Co. 7, 77, 79
Society of Motor Manufacturers and
    Traders 11, 42, 52, 120
'sourcing' 109–10
specialist car manufacturers 75, 94,
    see under corporate names,
Standard Motor Co. 7, 30–9, 56, 63,
    75, 77, 79, 84, 94, 98
standardization 47
sterling 47, 56–7
Stock Exchange 27
Stokes, Donald 85–7, 90, 100
strikes 25, 65–70
structure
    corporate 102
    industrial 10–11, 37, 41, 76–7
styling 33

Talbot 113
tariffs 12,19, 47, 95–6
    McKenna 7, 9, 11, 14
tax
    horsepower 14–17, 34, 52–3
    petrol 16
    purchase 56, 58
technical innovation 4–5, 32–3, 38,
    61, 96, 98–9
trade unions 8, 23–4, 43, 62, 64–5,
    116
training 82
Transport and General Workers'
    Union 63–4
Triumph Motor Co. 75, 77, 79

Vauxhall Motor Co. 16, 23, 25, 39,
    62, 71, 79, 91–2, 95, 110, 112–
    13, 122
Volkswagen 18, 53–4, 57–8, 97, 108,
    116–17

war 7–9, 43, 53, 73
Wolseley Motor Co. 5, 8–9, 28, 40,
    77, 78